IGNORANCE

BOOKS BY GEORGE G. SZPIRO

Kepler's Conjecture: How Some of the Greatest Minds in History Helped Solve One of the Oldest Math Problems in the World (Wiley, 2003)

The Secret Life of Numbers: 50 Easy Pieces on How Mathematicians Work and Think (Joseph Henry Press, National Academy of Sciences, 2006)

Poincaré's Prize: The Hundred-Year Quest to Solve One of Math's Greatest Puzzles (Penguin, 2007)

Numbers Rule: The Vexing Mathematics of Democracy, from Plato to the Present (Princeton University Press, 2010)

A Mathematical Medley: Fifty Easy Pieces on Mathematics (American Mathematical Society, 2010)

Pricing the Future: Finance, Physics, and the 300-Year Journey to the Black-Scholes Equation (Basic Books, 2011)

Risk, Choice, and Uncertainty: Three Centuries of Economic Decision-Making (Columbia University Press, 2020)

Perplexing Paradoxes: Unraveling Enigmas in the World Around Us (Columbia University Press, 2024)

Random Numbers Unveiled: The Secrets of Numbers That You Can't Predict but Can Rely On (Taylor & Francis, 2026)

The three wise monkeys: "Hear no evil, speak no evil, see no evil."
From Wikimedia commons: https://commons.wikimedia.org/wiki/
File:Toshogu_Three_Monkeys_Restored_2024_Crop.jpg

IGNORANCE

**WHAT WE DO NOT KNOW,
CANNOT KNOW,
MUST NOT KNOW,
AND REFUSE TO KNOW**

GEORGE G. SZPIRO

Columbia University Press   *New York*

Columbia University Press
Publishers Since 1893
New York   Chichester, West Sussex
cup.columbia.edu

© 2026 George G. Szpiro

Library of Congress Cataloging-in-Publication Data
Names: Szpiro, George, 1950– author
Title: Ignorance : what we do not know, cannot know, must not
know, and refuse to know / George G. Szpiro.
Description: 1. | New York : Columbia University Press, [2025] |
Includes bibliographical references and index.
Identifiers: LCCN 2025032595 | ISBN 9780231221658 hardback |
ISBN 9780231563833 epub | ISBN 9780231565332 (PDF)
Subjects: LCSH: Ignorance (Theory of knowledge)
Classification: LCC BD221 .S97 2025
LC record available at https://lccn.loc.gov/2025032595

Cover design: Noah Arlow
Cover image: Shutterstock

GPSR Authorized Representative: Easy Access System Europe,
Mustamäe tee 50, 10621 Tallinn, Estonia, gpsr.requests@easproject.com

Dedicated to the generation after the next:

Tamar, Rotem, Daniel, Ruben, Romy, Lavie, Maya, and Gali

(Status June, 2025)

Knowledge is a big subject. Ignorance is bigger. And it is more interesting.

—Stuart Firestein, *Ignorance: How It Drives Science* (2012)

CONTENTS

II CANNOT KNOW

III MUST NOT KNOW

IV REFUSE TO KNOW

ACKNOWLEDGMENTS

Jonathan Burke and Eva Burke-Szpiro deserve sincere thanks for reading the entire manuscript and offering numerous suggestions that have been incorporated.

I also extend my appreciation to Brian C. Smith, editor at Columbia University Press. This is the third book of mine that he accompanied, patiently, purposefully, and professionally, from when it was still just a concept all the way to publication. Thanks are due to several anonymous reviewers who were kind enough to examine the text closely and make judicious comments.

Finally, uxorial gratitude goes to my wife Fortunée, filial appreciation to my children and their spouses, and nepotistic pride and joy to an expanding number of grandchildren, all of whom constantly inspire me to reflect and to come up with new ideas.

IGNORANCE

INTRODUCTION

See No Evil, Hear No Evil, Speak No Evil

Aseventeenth-century carving over a door of the Tōshō-gū shrine in Nikkō, Japan, depicts three monkeys: Mizaru, covering his eyes, Kikazaru, covering his ears, and Iwazaru, covering his mouth. The carving symbolizes the Japanese principle "see no evil, hear no evil, speak no evil." The iconic image of the three monkeys has been reproduced many times and in many different contexts. Even Mahatma Gandhi, who eschewed most worldly possessions, owned a replica.

The carving was probably inspired by a maxim of the Chinese sage Confucius (551–479 BCE), whose sayings were collected by his followers and published many years after his death as the "Sayings of Confucius," aka *The Analects*. In Book XII Confucius's disciple Yen Yuang asked the master how one should practice perfect virtue. "Look not at what is contrary to propriety; listen not to what is contrary to propriety; speak not what is contrary to propriety," Confucius answered. In some portrayals a fourth monkey is added; he covers his genitals to indicate that one is to "do no evil."

The motto—not to see, not to hear, not to speak—is considered a wise dictum in the Far East, indicating noble conduct, but this is not so in the Western world. To turn a blind eye toward evil, to close one's ears to wickedness, or to fail to speak out against misdeeds suggests a lack of moral responsibility. People who refuse to acknowledge impropriety are not considered noble but rather the opposite. Feigning ignorance by looking the other way indicates spinelessness or, worse still, complicity.

The lack of courage to speak out calls to mind the code of *Omertà*, the conspiracy of silence of Mafia-type criminal organizations. Accordingly, the response of the above-mentioned Yen Yuan to the master's exhortation befits a member of a street gang rather than an adherent to Confucian values: "Though I am deficient in intelligence and vigor, I will make it my business to practice this lesson."

There is another facet to "See Not, Hear Not, Speak Not." Overlooking everything by closing one's eyes or disregarding everything by plugging one's ears means ignoring not only the evil but also the good; keeping one's mouth shut means not only refraining from slanderous rumors but also ignoring the praiseworthy.

ཨ

Epistemology, the theory of knowledge, is an intellectual discipline that has been discussed, dissected, and debated for thousands of years. Ignorance, on the other hand—its nature, causes, and implications—has received relatively little systematic attention as a theory, despite its profound influence on human behavior, societal structures, and—somewhat paradoxically—even on the pursuit of knowledge itself. It is true that Socrates, Plato, and Aristotle, and myriad thinkers who followed them, discussed ignorance, but mainly as a contrast to their primary interest: the pursuit of knowledge.

Only in modern times has ignorance emerged as an area of scholarly study in its own right.[1] If knowledge was defined as *true belief* by Socrates in the fifth century BCE, or as *justified true belief* by Edmund Gettier in the twentieth century, today ignorance is considered not just as the absence of belief but as a phenomenon shaped by societal, political, psychological, and even scientific and mathematical factors. Reflecting the limits of human understanding, the inherent constraints of nature and logic, the prohibitions against what must remain undisclosed, and the inner resistance to what is unwelcome, ignorance arises from what one does not know, cannot know, must not know, or refuses to know.

※

Generally, humankind's efforts are directed toward the accumulation of knowledge through education, research, observation, innovation, and experimentation. Whereas knowledge is actively produced and shared, ignorance results from neglect, systemic barriers, or deliberate efforts. Although generally deemed a negative state, ignorance does have a redeeming feature: as long as it is acknowledged, it motivates curiosity and drives the quest for knowledge.

One could suppose that as more and more knowledge is accumulated ignorance should diminish. Paradoxically, the opposite is true. In this, ignorance may be likened to the expanding universe. The more that cosmologists explore it, the more they realize how vast the as yet unseen—and expanding—parts are. Similarly, each new piece of knowledge unveils new pockets of ignorance and raises additional questions. For this reason, ignorance should not be viewed as a static void but as a dynamic, ever-expanding expanse.

※

In this book I classify concepts, ideas, and phenomena about which we are ignorant into four categories: what we do not know, cannot know, must not know, and refuse to know. The demarcations are not strict, and the choice to include a subject in this or that category may at times be arbitrary. Some chapters could fit into multiple categories.

I

DO NOT KNOW

I n this part of the book, we explore subjects that are unknown and questions that are unanswered. Some of them may become known and answered in the future if scientists and mathematicians keep looking hard enough. Ignorance about other matters may be self-imposed; still others remain unknown due to an inherent lack of self-awareness. Finally, some puzzling problems may never be solved because philosophers cannot agree on fundamental principles.

1

DO ALIENS VISIT US?

Flying Saucers and Other UFOs

Theories about unidentified flying objects (UFOs) have been something of a playpen for conspiracy enthusiasts, paranoia peddlers, rumor mongers, alternative thinkers, truthers, deniers, and sundry fringe hobbyists since the late 1940s. Photographs of flying saucers, accounts of sightings, reports about visitors from outer space, and stories of abductions have filled the tabloids.

Although most dismiss UFOs, the US Air Force took it upon itself to investigate UFO reports. *Project Sign* in 1948 was followed by *Projects Grudge* and *Twinkle* in 1949 and *Project Blue Book* (1952–1969). Other countries undertook similar investigations. According to the official announcements, no evidence of alien visitations or extraterrestrial visitations was ever found. The typically fuzzy and shaky sightings, captured by sensors or cameras that were not designed or calibrated for this purpose, were explained as the result of natural phenomena, faulty technology, delusions, and practical jokes. Indeed, after further investigations, most alleged flying objects turned out to be banal occurrences that were misidentified by amateur astronomers or impressionable observers: planets or comets or bright stars, military or civilian aircraft, surveillance or weather balloons, unusual cloud formations, atmospheric occurrences, light phenomena, and even psychological hallucinations or hoaxes.

Doubts remain, however. Practically all large-scale studies were undertaken by military authorities, and as is the wont of secrecy-minded defense officials, most details, except for the largely meaningless conclusions, remained classified. Due to distrust in government candidness, and spurred on by books, films, television, and social media, it is no wonder that four out of ten Americans still believe that some of the UFOs are alien spacecraft from other planets or galaxies, according to a 2021 Gallup poll.

༃

Belief in the existence of extraterrestrial beings is not all that is misplaced. Even though the probability of discovering aliens, or of their visiting us, may be infinitely small, the infinite vastness of the universe makes it somewhat presumptuous to assume that we earthlings are the only intelligent beings in the cosmos. To dismiss the possibility out of hand, as many "rational" scientists, afraid of being labeled pseudoscientists, would have us believe, is itself unscientific.

To remove the conversation about UFOs from the exclusive and secretive purview of military and national security officials, and to shift it from sensationalism to science, in 2022 the National Aeronautics and Space Administration (NASA) announced an independent study of what the US government now terms "unidentified anomalous phenomena" (UAP). A sixteen-member body was formed with leading experts from scientific fields, ranging from physics to astrobiology, to guide the direction of the study group.

Not to be left behind, after decades spent deflecting, debunking, and discrediting reports of unidentified flying objects, the

Pentagon formed the All-domain Anomaly Resolution Office (AARO) to address questions about UAPs and to mitigate the potential threats to safety and security posed by them. Using a rigorous framework and a data-driven approach, US service members and civil aviators are encouraged to report sightings that are then analyzed by defense and intelligence professionals.

Although the existence of intelligent alien life has not been ruled out, so far both NASA and defense-intelligence officials have stressed that no evidence has been found that would suggest an extraterrestrial origin for UAP sightings.

୬ଛ

The search for extraterrestrial intelligence (SETI) has spawned numerous initiatives outside of NASA and the Pentagon: for example, by the SETI Institute in Mountain View, California, or the Berkeley SETI Research Center. Both SETI research and UAP analyses investigate whether alien beings exist and are trying to visit or contact us, but their approaches differ. UAP projects interpret serendipitous chance observations, whereas SETI conducts deliberately planned searches. Examining data from ground and space-based radio telescopes and radio interferometers, SETI projects look for nonnatural radio or light signals that come from deep space.

To aid the analysis of the massive amounts of data with the limited resources at its disposal, the Berkeley SETI Research Center initiated a crowdsourced computing project named SETI@home. Volunteers from all over the world donated idle power of their laptops and home computers to scan through the data from radio telescopes and look for possible signs of extraterrestrial intelligence. For two decades, from 1999 to 2020, the crowds' personal computers and laptops searched for signals that stood out from random noise. SETI@home ended its crowd-sourced search in 2020; the organizing team has since focused on analyzing the accumulated data. To date, no extraterrestrial signals have been identified.

ཟ▲

One researcher who is ready to stake his illustrious academic career on the search for signs of extraterrestrial intelligence is Avi Loeb, an Israeli-American astrophysicist at Harvard University. He cannot be labeled a pseudoscientist because he has all the right credentials: he is a tenured professor at a first-rate university with hundreds of scientific papers in peer-reviewed journals, the former chairman of Harvard's astronomy department, a member of learned societies, and the recipient of numerous honors and awards. In addition to being a distinguished scientist, Loeb has a knack for public relations. With financial help from several enthusiastic billionaires, he founded the Galileo Project for the Systematic Scientific Search for Evidence of Extraterrestrial Technological Artifacts. The project's aim is not only to identify and analyze electromagnetic signals but to search for physical objects—actual items that can be looked at, touched, and felt—that could possibly stem from extraterrestrial technological equipment.

Pivotal for his thinking was the discovery in 2017 of Oumuamua, the first interstellar object that passed through the solar system. Loeb speculates that this object—which is now traveling away from us again at a speed of twenty-six kilometers per second—might actually be a probe from an intelligent species from outer space that is trying to learn about our solar system. According to his theory, Oumuamua would be a kind of alien spacecraft propelled by a light sail, a super thin

Oumuamua (artist's impression)

metallic sheet that catches sunlight in the same way a ship's sail catches the wind. If Loeb is right, which is still very, very, very questionable, the object would not be a natural phenomenon but an artifact, manufactured by extraterrestrial aliens. A feasibility study is underway about building a spacecraft that would be able to catch up with Oumuamua and have a closer look at it in about twenty-six years.

In 2023, Loeb organized an expedition to Papua, New Guinea, where an unusual meteorite had fallen into the ocean a decade earlier. He had the ocean floor swept with specially designed machinery that found magnetic fragments. Loeb claimed that they were parts of material that belonged to a big object that must have come from outside the solar system. Proof of an extra-terrestrial alien-made spacecraft? Loeb's peers are skeptical.

2

KNOWLEDGE UNDER SEAL

Les Plis Cachetés

Every scientist strives to be the first to establish a hitherto unknown scientific fact, to prove a mathematical theorem, to invent a new machine, to discover a new species, or to develop a new drug. The quest for priority is what drives many scientists to spend days and nights on their scientific enterprise.

Unfortunately, disagreements about priority often lead to disputes that can sometime turn ugly. In the late seventeenth and early eighteenth century, the English natural philosopher Isaac Newton and his German counterpart Gottfried Wilhelm Leibniz engaged in an acrimonious dispute about who had invented the infinitesimal calculus. Awash with nationalistic overtones, they and their followers accused the other side of plagiarism, although modern historians of science are convinced that they both developed the infinitesimal calculus independently. And the wish to quickly establish priority may have led Pierre de Fermat to write a hurried note on the margin of a book. Had he taken the time to write down and publish the supposedly concise proof of his eponymous theorem, he would have saved hundreds of mathematicians many years of hard labor.

Actually, the quest for priority has an important and decidedly positive aspect. If priority were of no concern, science would advance at an unhurried pace and new research results would be circulated in a very leisurely manner. It is the desire to be first that provides the incentive for scientists to work tirelessly on their projects and thus make rapid advances possible.

But as beneficial as this desire is for science and society, why do individuals themselves subscribe to it? Why do researchers want the world to know that they were the first to make a discovery? Recognition and reputation that come with being a forerunner are the principal motives. Immortality—by having a theory, a theorem, an elementary particle, a plant, an animal, an island, or a lunar crater named after them—is another. An additional reason is that pioneers are able to inspire others and shape the direction of future research. Of course, there are also financial rewards for the first person to state a scientific finding, through fellowships, professorships, commercial ventures, and prizes that encourage the progress of science (for example, the million-dollar Nobel Prizes, the Clay Prizes, the Genesis Prize, and others). Last but not least, there is the personal satisfaction for a scholar to know that he or she discovered something that nobody else was aware of yet.

With so many reasons to announce one's discovery to the world as fast as possible, would anyone consider hiding one's findings? Would any serious scientist want to keep the scientific community ignorant of a discovery?

Apparently yes. The *Académie des sciences* (the French Academy of Sciences, founded in 1666, then named the *Académie royale des sciences*) instituted a policy in 1735 that allows any person to deposit a sealed envelope, a *pli cacheté*, in its archive that the academy would keep secret. In it the depositor may have put either a paper containing his ideas or an object of her invention. (Today the size of the envelope is limited to 38 × 25 × 1.5 centimeters.) Upon deposition of the envelope, the date is noted and the envelope is kept secret for a minimum of two years. After that time, the depositor or, if deceased, the heirs may demand the opening under the auspices of the *Commission des plis cachetés.* The commission takes note of the subject matter and after consultation decides whether to publish the contents. The academy reserves the right to open any as yet unopened *pli*s after one hundred years. In all cases, the original content remains in the academy's archives.

Why might someone want to keep an idea or an invention or a discovery secret in this way? One reason is that the sealed envelope method ensures that the depositor would one day, upon opening of the pli, be recognized as the first to have made that particular scientific contribution. But that could have been achieved by publishing the contribution, so why bother with the pli?

There are reasons why authors may not want to share their findings publicly but nevertheless reserve the right to be recognized one day as the progenitors. For example, the idea may not yet be fully developed; by depositing the envelope, even if the work is not yet complete or fully verified, the researcher can continue to refine it or gather additional evidence without risking that someone else would be able to claim priority. Or he or she may need to secure additional funding or resources to pursue further research. Or the findings may not be politically correct at the time they are made, and the depositor may have concerns about potential implications, backlash, or criticism. Some authors may want to wait for the right time to publicly share their work, such as when it would have the greatest impact, align with a planned conference, or coincide with a favorable political or social climate.

Today researchers are usually not coy about their discoveries, and secrecy is usually no longer of any concern. Manuscripts are submitted to preprint servers, preliminary results are presented at scientific conferences, and patents are filed. Many half-baked ideas find their way to the public.

≈

One of the most intriguing sealed envelopes was sent to the academy in early 1940 from the front lines of the battles of the Second World War by a soldier by the name of Wolfgang Döblin.[1] Son of the famous German writer Alfred Döblin, his family had left Berlin in 1938. They were of Jewish origin, and even though they considered themselves Protestant, they did not feel safe in Nazi Germany. In Paris where they had found refuge, Wofgang, now Vincent, quickly made a name for himself as a brilliant student of mathematics. By the time he was twenty-three, he had already published eight papers and another nine announcements in the *Comptes Rendus de l'Académie des Sciences*. A rising star, he was destined for a brilliant academic career. But he insisted on enlisting in the French army and serving as a simple soldier. In this manner, he told his brothers, he could show his gratitude to a country that had accepted him and his family at a time of desperation.

Serving as a radio operator at the front, he continued his mathematical research. But the French battle lines were failing, and as the Germans drew near he knew that as a Jew and as the son of the decadent socialist writer Alfred Döblin his fate would be sealed if he were to be caught. On the morning that the German army invaded the village in which he had found refuge, Vincent burned all his papers and shot himself.

Unbeknown to anyone, four months before his death, Döblin had sent a *pli cacheté* to the academy. A single reference to the *pli* was contained in one of his letters from the front. It was only in 1991, half a century after Döblin's death, that Bernard Bru, a historian of science, discovered that letter. Bru contacted Döblin's

FUYANT LE NAZISME
L'ÉCRIVAIN ALLEMAND
ALFRED DÖBLIN
1878 - 1957
S'INSTALLA DANS CET IMMEUBLE
AVEC SA FAMILLE DE 1934 À 1939.

POURCHASSÉ PAR LA WEHRMACHT
SON FILS WOLFGANG DÖBLIN
1915 - 1940
MATHÉMATICIEN
PRÉCURSEUR DU CALCUL DES
PROBABILITÉS
EST MORT POUR LA FRANCE
À HOUSSERAS (VOSGES)
LE 21 JUIN 1940
À L'ÂGE DE 25 ANS.
TITULAIRE
DE LA MÉDAILLE MILITAIRE
ET DE LA CROIX DE GUERRE

Memorial plaque at the Döblin family's home in Paris

surviving brother, who agreed to the opening of the sealed envelope. On May 18, 2000, the secret was lifted.

And what a surprise it was! The subject treated in Döblin's booklet was Brownian motion, the random movement of a particle in a liquid medium. Döblin had discovered extremely important properties of so-called stochastic processes that were rediscovered only some twenty years later by the Japanese mathematician Kiyoshi Itō and others, generally under more restrictive hypotheses.

When Itō was ceremoniously inducted into the French Academy of Sciences in 1989, the *pli cacheté* still rested, its seal unbroken, in the dusty catacombs of the same institution. Itō had trodden the paths of stochastic theory, ignorant of the fact that these same paths had been broken long before by the unknown French soldier who had preferred to die by his own hand rather than fall into the hands of the enemy.

3

A MATHEMATICAL RIDDLE

The Riemann Hypothesis

For several years Edge.org, a lively website for academics and intellectuals interested in science and technology, posed an annual question that would stimulate the entire community. In 2013 the annual question was: "What should we be worried about?" In all, 155 responses were received, but unfortunately I missed the deadline, a slip that I still regret because I was, and still am, very worried about a mathematical riddle.

The riddle is the so-called Riemann Hypothesis, named after the German mathematician Bernhard Riemann (1826–1866). It is a question about the Zeta function, specifically, about where in the xy-plane of complex numbers the function becomes zero. The so-called trivial zeros lie along the x-axis at –2, –4, –6, . . . It is the set of nontrivial zeros, however, the zeros that do not lie along the x-axis, that puzzled Riemann. Although he had a hunch about them, he was unable to prove it. He could only claim that it is *"sehr wahrscheinlich"* (very likely) that all nontrivial zeros of the Zeta function have a real part equal to ½. According to his hunch, it is "very likely" that all nontrivial zeros lie on this critical line, the vertical line along $x = ½$.

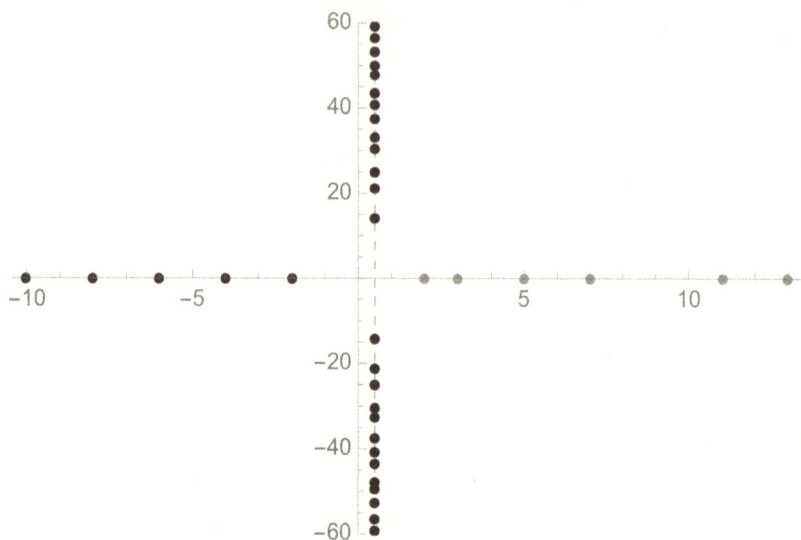

Trivial zeros (on the left), nontrivial zeros (on the vertical),
and prime numbers (on the right)

This hypothesis still remains no more than a conjecture; mathematicians are ignorant about whether Riemann's hunch is correct. The vast majority believe the hypothesis to be true, but nobody knows for sure. And the worry that keeps me up at night (well, not every night) is that it could turn out to be false.

Why worry? Many mathematical hypotheses exist and, apart from the people specifically working on them, hardly anybody loses any sleep over them. But the Riemann Hypothesis is in a class of its own. It is important: very, very important! In fact, a one million dollar prize has been offered by the Clay Foundation to anybody who proves or disproves it. The reason for the importance of Riemann's Hypothesis is that there is a subtle connection of the hypothesis to the distribution of prime numbers, the building blocks of our number system. Among other things, the Zeta function's nontrivial zeros can be used to describe how prime numbers are distributed among the integers. Therefore, the Riemann Hypothesis has profound consequences for our understanding of numbers.

One easy way to disprove a hypothesis is to find a counterexample. All it would take to blow the Riemann Hypothesis asunder is one nontrivial zero that *does not* lie on the critical line. So starting in 1859, the year when Riemann proposed his hypothesis, the race was on to find such a counterexample. In 1903, the Danish actuary J. P. Gram computed the first fifteen zeros; they all lay on the critical line. In 1956, mathematicians had extended the search to fifteen thousand zeros, by 1966, to a quarter of a million, by 1977 to forty million, and people were just getting started. Using the idle times of thousands of laptops and personal computers of volunteers around the world, the search for a counterexample was extended to two hundred million in 1982, to ten billion in 2001; by 2004, ten trillion zeros (10^{13}) had been verified. They all lay on the critical line!

Where does this leave us in terms of proving the Riemann Hypothesis? Precisely nowhere, because although one counterexample suffices to disprove a conjecture, even a googol (10^{100}) of supportive examples do not a proof make.

With no counterexamples in sight, efforts are underway, and have been for one and a half centuries, to prove that the nontrivial zeros *must* lie on the critical line. As of this writing, there are no takers for the million-dollar prize and the honor that would go with a successful, watertight proof.

And why does that keep me up some nights? As I pointed out, Riemann's Hypothesis has deep implications for the distribution of prime numbers. And prime numbers are the basic building blocks of number theory, somewhat akin to atoms in chemistry or cells in biology. So, yes, the Riemann Hypothesis is very important.

It is therefore not surprising that thousands of papers depend on the hypothesis being true. Hundreds of mathematical theorems begin with a sentence such as "under the assumption that the Riemann Hypothesis is true, *xyz* is true." If the hypothesis is false, *xyz* and all such theorems will be wrong . . . unless they can be proved by other means. It would be a real paradigm change. As a mathematician once said: "If [the Riemann Hypothesis is] not true, then the world is a very different place. The whole structure of integers and prime numbers would be very different to what we could imagine. . . . it would be a disaster because we've built so much around assuming its truth."[1] On the other hand, this opens a different avenue for disproving the Riemann Hypothesis. From "under the assumption that the Riemann Hypothesis is true, *xyz* is true" it follows by contraposition that "if *xyz* can be shown to be false, the Riemann Hypothesis must be false."

The implications of the Riemann Hypothesis span many areas beyond mathematics: the hypothesis has consequences even for electromagnetics and quantum mechanics; the efficiency or correctness of some algorithms depend on the Riemann Hypothesis being true; and, of course, encryption, as we use it for example on the internet, depends crucially on prime numbers and hence on the correctness of the Riemann Hypothesis.

For now, and for the foreseeable future, we remain ignorant of the truth of the Riemann Hypothesis.

APPENDIX

The Riemann Hypothesis deals with the Zeta function,

$$\zeta(s) = \sum_{n=1}^{\infty} \frac{1}{n^s} = \frac{1}{1^s} + \frac{1}{2^s} + \frac{1}{3^s} + \dots, \text{check the zeta-function!}$$

where s is a complex number, made up of a real part and an imaginary part: $s = a + bi$. In the xy-plane, a-values (the real parts) are plotted along the x-axis, b-values (the imaginary parts) along the y-axis. It is obvious, to mathematicians at least, that the Zeta function is equal to zero when $b=0$ and $a= -2, -4, -6, \dots$. These are the trivial zeros. But there are infinitely more nontrivial zeros, with positive real parts. The Riemann Hypothesis says that the real parts of all nontrivial zeros are exactly equal to ½.

4

RESTRICTED ACCESS

Top Secret vs. Freedom of Information

"**P**residential Libraries and Museums promote understanding of the presidency and the American experience. We preserve and provide access to historical materials, support research, and create interactive programs and exhibits that educate and inspire." So says the website of the National Archives, the depository of governmental and historical records of the United States.

The fifteen Presidential Libraries currently in existence in the United States "hold the raw materials of history: evidence of democracy at work and of the continuing relevance of the Presidents' past decisions." They are not traditional libraries but are repositories for preserving and making accessible the papers, records, and other historical materials of US presidents. The Presidential Records Act (PRA), enacted in 1978, allows public access to records of a president's administration five years after the end of that administration.

Can we, therefore, learn all about what went on in the Oval Office? Well, not quite! Access to historical material, in particular about the goings-on in the Oval Office, is not unfettered. There are limitations to what and when the public is permitted to know.

The American president may invoke a right to restrict public access to certain information and documents for up to twelve years. The specific restrictions relate to national security, foreign policy, appointments to federal office, certain trade secrets, certain commercial or financial information, confidential communications between the president and the president's advisers, and medical files. The president's restrictions are only overridden if the information is necessary to an incumbent president, to the House of Congress, or to a committee or subcommittee in order to conduct government business, or if a court has issued a subpoena for the purpose of a civil or criminal investigation or proceeding. Then the documents must be produced.

In turn, the overrides can also be overridden, and some categories of information will remain out of reach. They include classified defense information, internal agency rules, trade secrets, protected communications, personnel files, law enforcement records, regulatory documents, and geological data. Records or information compiled for law enforcement purposes can be withheld, however, only when it

> could reasonably be expected to interfere with enforcement proceedings . . . would deprive a person of a right to a fair trial or an impartial adjudication . . . could reasonably be expected to constitute an unwarranted invasion of personal privacy . . . could reasonably be expected to disclose the identity of a confidential source . . . would disclose techniques and procedures for law enforcement investigations or prosecutions, or would disclose guidelines for law enforcement investigations or prosecutions if such disclosure could reasonably be expected to risk circumvention of the law . . . or could reasonably be expected to endanger the life or physical safety of any individual.[1]

The United States, like most countries, has three levels of classification:

> Top Secret shall be applied to information, the unauthorized disclosure of which reasonably could be expected to cause exceptionally grave damage to the national security that the original classification authority is able to identify or describe.
>
> Secret shall be applied to information, the unauthorized disclosure of which reasonably could be expected to cause serious damage to the national security that the original classification authority is able to identify or describe.
>
> Confidential shall be applied to information, the unauthorized disclosure of which reasonably could be expected to cause damage to the national security that the original classification authority is able to identify or describe.[2]

In the United States, the Top Secret classification may be applied only by the president, the vice president, an agency head, or an official designated by the president. Authority to classify information as Secret or Confidential may be delegated to other high-ranking officials.

Access to classified information is restricted on a need-to-know basis to people who have been determined by an agency head to be eligible. There are many people who need to know: as of October 2019, 1.7 million American government employees and contractors had a "confidential" or "secret" clearance, and another 1.2 million had a "top secret" clearance.[2] That corresponds to more than 1 percent of the adult population of the United States.

The kinds of information that are classified include military plans, weapons systems, and operations; foreign government information; intelligence; foreign relations or foreign

Top secret stamp

activities; scientific, technological, and economic matters relating to national security; programs for safeguarding nuclear materials or facilities; vulnerabilities or capabilities relating to national security; and information about the development, production, or use of weapons of mass destruction.

There is one more level of secrecy, although it is not strictly a classification: Sensitive Compartmentalized Information. This designation can apply both to Secret and to Top Secret documents and indicates that the document cannot be viewed by anybody, even if he or she has the appropriate clearance. Ony a limited number of specially authorized people are allowed to view documents classified as TS/SCI (Top Secret/Sensitive Compartmentalized Information).

៛៛

For a democracy to function, people must have the right to know the business of their government. After all, transparency is one of the main features that distinguishes a democracy from authoritarian regimes. Hence, like many free countries, the United States has a Freedom of Information Act (FOIA). Originally enacted in 1966, FOIA requires federal agencies to disclose unreleased information and documents if it is requested by a citizen, a business corporation, a law firm, or a news organization. (Individual states and local governments have similar legislation.) Even interagency or intra-agency memoranda or letters can only be withheld for twenty-five years.

Hence, as important as it is to keep the public ignorant about confidential, secret, and top secret information, it is equally important to keep the public as informed as possible about governmental activities. On December 29, 2009, President Barack Obama signed Executive Order 13526, which prescribes a uniform system for classifying, safeguarding, and declassifying national security information. Significantly, however, it also specified that no information may remain classified indefinitely. The order created a National Declassification Center whose task it is to ensure that information that does not require protection in the interest of the national security not be classified. (The shortcut term "classified" is a bit of a misnomer. One can classify information according to many characteristics: secret vs. nonsecret is just one classification scheme.)

The decision whether or not to restrict certain information is not always easy. The president, the officials, and the institutions responsible for providing or restricting information must navigate between the two requirements: on one hand, it is essential that accurate information be made available to the people for them to hold the government accountable; on the other hand, intelligence critical to the nation's security must be protected. Thus at the same time as classification standards are rigorously applied, the relevant authorities must institute effective and secure routines to declassify information that no longer requires secrecy. (Presumably, President Trump's claim that he can declassify items simply by thinking about them would not be considered an effective and secure routine.)

After ten years information must be automatically declassified unless the original classification authority determines that the sensitivity of the material requires that it remain secret for up to twenty-five years due to concerns involving information on intelligence sources or methods, weapons of mass destruction, US cryptologic and military systems, war plans, foreign relations, national security preparedness, protection of high officials, or legal restrictions under statutes or agreements. Classified records that are more than twenty-five years old and have been determined to have permanent historical value shall, with certain exceptions, be automatically declassified.

Finally, some information can be kept secret for seventy-five years or more if the Interagency Security Classification Appeals Panel—made up of representatives of the Departments of State, Defense, and Justice, the National Archives, the Office of the Director of National Intelligence, and the National Security Advisor—so decides. Unfortunately, the specific types of information that are exempted from public release under Executive Order 13526 for seventy-five years are not listed. Why?

Apparently, even the information of what is to be kept secret is secret.

Under no circumstances may information be classified in order to conceal violations of law, inefficiency, or administrative error; prevent embarrassment to a person, organization, or agency; or to restrain competition. Neither shall basic scientific research be classified unless it is clearly related to national security.

5

IGNORAMUS ET IGNORABIMUS

A Controversy in Germany

Gentlemen!

Just as a world conqueror of olden time, on a day of rest in the midst of his victorious marches, could desire to see the borders of the immense stretches of land subjected to his rule determined more precisely—here in order to draw tribute from a people not yet subjected to taxes, there to recognize a natural obstacle in the water desert, insurmountable to his horsemen—thus to recognize the true limit of his power: so it will not be an inappropriate beginning for the world champion of our days, the natural science, if she tries, resting from her work on a festive occasion, to clearly delineate the true borders of her immeasurable empire.[1]

Thus began a lecture by the German physiologist Emil du Bois-Reymond (1818–1896) to an audience of distinguished scientists and medical doctors, the Society of German Natural Scientists and Physicians (*Gesellschaft der Deutschen Naturforscher und Ärzte*), in Leipzig on August 14, 1872. The theme the world-famous physician intended to discuss, expressed concisely in four words instead of one-hundred-and-fifteen, was "the limits to science."

Du Bois-Reymond had done pathbreaking investigations of electrical properties of the nervous system. Nevertheless, his outlook was pessimistic. Convinced that some fundamental questions, in particular, the nature of matter and force and the relationship between mental phenomena and their physical aspects, would forever remain beyond the reach of human understanding, he closed the speech on a gloomy note. With respect to physiology, Du Bois-Reymond concluded that the scientist has long ago become accustomed "in manly resignation" to the concept of *ignoramus* (Latin for "we do not know"). Nonetheless, considering the many past achievements, one may remain hopeful that even what is not known today may become known some day. Concerning the riddles of matter and force, however, and human consciousness, scientists must once and for all come to terms with the much harder truth that they will never know. He ended his lecture with the ominous exclamation "*Ignorabimus!*" (We will forever remain ignorant!).

The speech excited considerable controversy in German academic circles. Under the heading *Ignorabimusstreit* (ignorabimus controversy), scientists argued about what was knowable about consciousness and physical existence. Eight years later, in July 1880, Du Bois-Reymond took to the podium again, this time in front of the German Academy of Sciences. He enumerated a number of World Riddles, three of which—the nature of matter and force, the origin of motion, and the origin of sensation—he deemed not to be explainable by science or by philosophy. This time he was more cautious, however; he ended this lecture with a less definite exclamation: "*Dubitemus*" (We might doubt).

The storm (maybe too strong a word, but by the staid academic standards at the time it truly was a storm) did not abate. The world's then leading mathematician David Hilbert (1862–1943) of Göttingen University was particularly intrigued.

Du Bois-Reymond

Emil du Bois-Reymond

In 1900, at the International Congress of Mathematicians in Paris, he proposed twenty-three problems that were of particular importance and had so far eluded proof or disproof. In stark contrast to Du Bois-Reymond's assertion, Hilbert was convinced that "every definite mathematical problem must necessarily be susceptible of an exact settlement. . . . This conviction of the solvability of every mathematical problem is a powerful incentive to the worker. We hear within us the perpetual call: There is the problem. Seek its solution. You can find it by pure reason, for in mathematics there is no *ignorabimus*."[2]

In 1912 the *Journal of Philosophy* published an appeal "to all who are interested in promoting the scientific spirit in philosophy." It was signed by Albert Einstein, the father of relativity theory, Sigmund Freud, the father of psychoanalysis, David Hilbert, Ernst Mach, the physicist/philosopher, and about thirty other distinguished German and Austrian luminaries.

> There has long been felt the need of a philosophy which should grow in a natural manner out of the facts and problems of natural science. The mechanical view of nature no longer satisfies this need. Let anyone recall the "*ignorabimus*" of Du Bois-Reymond. . . . The present day desires the solution of general problems, which research itself throws up, and is not to be put off with an ignorabimus for which there is no evidence.[3]

On the occasion of his retirement in 1930, Hilbert gave a radio address in which he reiterated his optimistic conviction that there is no *ignorabimus*. His talk ended with the six words that were inscribed, thirteen years later, on his gravestone: "*Wir müssen wissen. Wir werden wissen*" (We must know. We shall know.).

Many of Hilbert's twenty-three problems have been answered in the first century following the speech. Some are still open, with people actively working in the hope and expectation that they will be solved in the second century after the famous speech. But one problem shot Hilbert's optimism to smithereens. Problem number two on the list sought a proof that the axioms of arithmetic are consistent. The surprising answer, given by a then still unknown Austrian logician Kurt Gödel (1906–1978) challenged Hilbert's *Wir werden wissen*.

Ironically, it had been just two days before Hilbert's speech, at a roundtable discussion on September 6, 1930, that Gödel presented a theorem that would become known as the "incompleteness theorem" (see chapter 27). The theorem says that in a formal system (like arithmetic) statements exist that are unprovable within that system.

This is much worse than a problem just being unsolvable. Hilbert's tenth problem, for example, asked whether there was an algorithm that could determine if a Diophantine equation is solvable. The answer, found in 1970 by the Russian mathematician Yuri Matiyasevich, is that no such algorithm exists. Hence we know that the answer to Hilbert's tenth question is "No!" Gödel's theorem, on the other hand, says that in any consistent mathematical system there are statements whose truth or falsity cannot be determined within that system.

For Hilbert's optimistic stance in the *ignorabimus* debate this was truly devastating. It showed that there are logical statements in formal systems, for example in arithmetic, that can neither be proved nor refuted with the axioms of that system; hence *ignoramus* and *ignorabimus*.

A positive aspect does, however, remain. To realize that a problem cannot be answered does not mean that we are ignorant. Gödel's incompleteness theorem is itself a most important form of knowledge: to know that something is undecidable means at least that we know we cannot know. To paraphrase David Hilbert in the original German: *Wir wissen* . . .

6

WILL THE UNIVERSE
EXPAND FOREVER?

Big Crunch, Big Bounce, Big Rip, or Big Chill

Our universe came into being with the Big Bang some 13.8 billion years ago—give or take a few million—and has been expanding ever since. A century ago, this was not yet known. Albert Einstein initially believed that the universe was in a steady state: always had been and always would be. This view was challenged, in 1927 by, of all people, a Catholic priest from Belgium, Georges Lemaître.

Lemaître had taken his vows at the Cathédrale Saint-Rombaut in Malines (Belgium) and his PhD at the Massachusetts Institute of Technology.[1] A first-class physicist and mathematician, he opposed mixing science with religion. He felt that science had its place in advancing our understanding of the natural world, whereas religion played a role in addressing questions related to the meaning of life, morality, and metaphysical matters.

His theory about the origin of the universe was published in a little-known Belgian journal.[2] When that theory (later named the Big Bang Theory) first came to Einstein's attention, he dismissed it. "Your calculations are correct but your physics is abominable," he told Lemaître after a lecture. But Einstein was soon disabused. In 1929, the American astronomer Edwin

Georges Lemaître

Hubble published his observation that distant galaxies were moving away from our galaxy at speeds proportional to their distance; the inference had to be that the universe was expanding. Following Hubble's reasoning in the other direction in time, one is forced to conclude that the universe must have originated in the Big Bang . . . just as Lemaître's equations suggested. Actually, the Soviet physicist Alexander Friedman (1888–1925) had already come to the same conclusion.

ॐ

Then in the late 1990s there was another shocker. In an attempt to measure the rate of expansion of the universe at different times in the past, two teams of astrophysicists observed distant supernovae. Their observations not only confirmed that the universe was expanding but that it was expanding at an accelerating rate. The reason for the acceleration was something called "dark energy." Nobody really knows what dark energy actually is; it is thought to be a theoretical form of energy that permeates all of space. In 2011, Saul Perlmutter, Brian Schmidt, and Adam Riess were awarded the Nobel Prize in physics for the discovery that the universe's expansion accelerates.

So the riddle of the past has been cracked, and the mystery of the present has been solved. But what about the future? Will the universe expand forever, or will it stop expanding and start to contract? And then what? Will it expand to infinity or contract to a singularity and then . . . poof . . . all is gone?

The simple answer is that we do not know. What scientists do seem to know is that there are two forces in the universe that act at cross purposes: gravitation pulls everything together and dark energy pushes everything asunder. If there was no dark energy, the universe would eventually stop expanding and start to contract due to gravity. But dark energy exerts a repulsive force that counteracts gravity and pushes everything apart. For as long as the universe expands, the density of matter decreases, and the repulsive force of dark energy eventually begins to overcome the gravitational attraction.

If contraction is what is in store, the universe will become increasingly dense and hot, eventually leading to a singularity, a point of infinite density and temperature, where the laws of physics as we know them break down. This "poof . . . all is gone" scenario is based on the assumption that the universe contains enough matter and energy so that gravity overcomes the expansion and lets everything collapse; this is popularly known as the Big Crunch.

But that may not be the end of it. The Big Crunch could be followed by another Big Bang, followed by another Big Crunch, followed by another Big Bang, followed by . . . you get the picture. Such a never-ending Big Bang-Big Crunch cycle has been called the Big Bounce.

However, if the universe continues to expand forever, the expansion will continue to accelerate and eventually become so fast that it will overcome the gravitational attraction of all matter and energy. Based on the assumption that the dark energy that drives the acceleration will continue to dominate the universe's evolution, the rapid expansion will tear apart even atoms and molecules, causing the universe to rip apart. Fittingly, this scenario has been given the name the Big Rip.

But if expansion continues forever, this is not all that will happen. As the universe continues to expand and, as a consequence, the density of matter and energy decreases, the rate of expansion may gradually slow down. Eventually, the universe will become so spread out that it will be dark, cold, and empty, with no stars, galaxies, or structures. It will cool down, eventually becoming too cold to sustain anything. Yes, this scenario has been named the Big Chill.

❧

The crucial variable, the key scientific question about the future of the universe, is the average density of matter. If the density in our universe is less than the so-called critical density—roughly five or six hydrogen atoms *per cubic meter—the universe will expand forever. If the density is greater than the critical density, it will eventually recollapse in a Big Crunch.* If the density is just equal to the critical density, the expansion will continue, but gradually slow down over time, never stopping completely. The universe will reach a state of equilibrium, where the gravitational attraction of matter and energy is precisely balanced by the expansion rate due to the dark energy. All matter and energy will be uniformly distributed and at the same temperature, making it impossible for any physical processes or life to exist. This scenario is also sometimes called the Big Chill because it represents a cold, dark, and featureless universe with no structure or complexity.

Although not impossible, the Big Crunch and the Big Bounce scenarios are considered unlikely because the acceleration of the universe points in the other direction. But we do not know what is in store for us because the exact value of the density is still subject to ongoing research and debate in the scientific community. Is it so sparse that dark energy will blow everything apart, is it somewhere near the critical density and everything will eventually die a slow death, or is it so large that gravity will cause everything to crash together?

Although we currently ignore the answer, the density appears to be tantalizingly close to the critical density. Hence, whether we look forward to the Big Rip or the Big Chill remains to be seen.

7

IGNORANTIA LEGIS NON EXCUSAT

Presumed to Know

Hugo took a bicycle that was parked in front of the supermarket to ride home. A passerby recognized the thief and informed the police. At court Hugo argued that he had not known it was illegal to borrow someone's bike without asking. He was sent to juvenile detention.

Dorothy traveled from Springfield, Missouri, to Springfield, Illinois. An avid Second Amendment advocate, she had a Glock pistol, legally acquired in her hometown, strapped to her hip. When she arrived at her destination, she was promptly arrested. At the hearing, Dorothy stated that it was legal to carry a gun in her Springfield and contended that she did not know that in this Springfield it was illegal without the proper paperwork. It was in vain; she was lucky to get away with a fine.

The argument in the first case is obviously without any merit whatsoever. The second case is a little trickier. Dorothy had never in her life traveled outside the state of Missouri and was not aware that laws in Illinois were different. However, the judge would not accept that argument: the law is the law! And one is supposed to be aware of it.

᠊ଈ᠊

The Latin dictum *"Ignorantia Legis Non Excusat"* (Ignorance of the law does not excuse) originated in Roman times, but the idea was understood and accepted much earlier. For five millennia, laws that bind all citizens to a common set of norms have been publicly displayed so nobody could claim ignorance (at least not those who were literate).

King Hammurabi, ruler of Babylonia in the third millennium BCE had his two hundred and fifty or so laws chiseled into a stone obelisk. In the seventh and sixth centuries BCE, the Greek lawgivers Draco and Solon had their laws inscribed on *Axones*, horizontally arranged wooden beams that could be rotated for all to see.

The Twelve Tables, the laws of the Roman Republic, dating from the fifth century BCE, were engraved on bronze tablets and publicly exhibited in the *Forum* in Rome. All Roman citizens had occasion to see, read, and study them; schoolchildren were obliged to learn them by heart. The Twelve Tables put an end to the previously unwritten laws, exclusively interpreted by upper-class priests, which had kept lower-class citizens in ignorance.

Moses, on the other hand, (ca. fourteenth century BCE) seems to have kept the two tablets on which God had inscribed the Ten Commandments hidden away from the public's eye, in the Ark of the Covenant. But the Israelites were supposed to know the laws nonetheless. The Bible, written in the fifth century BCE, clearly states (Leviticus 5:17) that "if a person sins and does what is forbidden in any of God's commands, even though he does not know it, he is guilty and will be held responsible."

In the sixth century CE, Emperor Justinian (482–565 CE) assembled a ten-man committee who compiled the laws, edicts, and decrees of previous rulers, going back four hundred years. The multivolume *Codex* eliminated obsolete material, cut unnecessary matter, and reconciled contradictions. It and its

Stele of Hammurabi

companion books form the basis of most legal systems around the world today.

It is in Book I, Title 18 of the *Codex* that the committee discussed cases concerning ignorance of law and of fact. Emperor Gordian, for example, had written to one of his subjects that because he was older than twenty-five, "you cannot readily be excused on account of your ignorance of the law." In a matter of inheritance, Emperor Philip informed a man who had neglected to claim his inheritance within the prescribed period that "you can, under no circumstances, allege ignorance of the law." And Emperor Diocletian, in answer to a petition by someone who had erroneously paid a debt, decreed that if "anyone, who is ignorant of the law, pays money which is not due, he cannot recover it; . . . only ignorance of fact confers the right to recover money which has been paid when it was not due."

ఇ.

In olden times, there were few laws and the literate among the people could read and memorize them. Today there are thousands of laws. Nevertheless, as the fictional examples at the beginning of the chapter indicate, ignorance is not accepted as an excuse. The law ignores ignorance.

The English jurist and scholar John Selden (1584–1654) said so in one of his *Table Talks* that were posthumously published in 1689: "*Ignorantia legis neminem excusat*" (Ignorance of the law excuses nobody). A century later, his compatriot, the jurist and judge Sir William Blackstone (1723–1780), phrased the principle a bit differently: "often a mistake in point of law which every person of discretion not only may, but is bound and presumed to know, is in criminal cases no sort of defence."[1] In 1761, the French jurist Robert Joseph Pothier (1699–1772) wrote: "no man shall, under the pretence of an ignorance of the law, excuse himself from the performance of his own obligations."[2] On the other side of the Atlantic, Justice Oliver Wendell Holmes (1841–1935), Associate Justice of the United States Supreme Court, wrote in 1881, in *The Common Law*, a book that is still in print one hundred and fifty years after its first publication, that "Ignorance of the law is no excuse for breaking it."[3]

Rarely, however, exceptions exist in which the regulations are so obscure that ignorance may serve as a defense. A famous example of when ignorance of a law served as a successful defense was the 1957 Supreme Court case Lambert v. California. A forger was ignorant of the fact that as a convicted felon she was supposed to register with the Chief of Police if she remained in Los Angeles for more than five days. She was convicted by a lower court even though she claimed she had been unaware of the ordinance. The Supreme Court, however, accepted her defense. Although the opposing justices held that "a mistake of law is usually not a defense to a crime. For example, a person may be convicted of drug possession even if he or she did not know that the particular substance was illegal," the majority decided that "the prosecution would need to provide circumstantial evidence showing that the defendant should have been aware of the possibility that registration was required."[4]

੨ֆ

To end, let us note that Blackstone's statement that every person is "presumed to know" the law is clearly mistaken. Ignorance is no excuse, but it cannot simply be presumed away. Sometimes decisions of District Courts are reversed due to errors in law by a High Court, whose decisions may, in turn, be reversed by the Supreme Court. Does that mean that the honorable judges were ignorant of the law? Well, yes, maybe. In any case, it is incorrect to simply presume that everybody knows the law.

8

WHAT IS NOTHING?

Horror Vacui

Question: What is a hole?

Answer: Nothing . . . surrounded by a border.

Question: Does Nothing exist?

Answer: Well, since holes exist, Nothing exists too.

Question: If Nothing exists, can nothing exist?

More generally: Why is there something rather than nothing?

The last question was pondered by the ancient Greek thinkers Plato and Aristotle. In modern times it has been raised by the German philosopher (and Nazi sympathizer) Martin Heidegger who deemed it one of the most important questions about the fundamental nature of reality. In his "Introduction to Metaphysics," a lecture course that was turned into a book, the first sentence reads: "Why is there something rather than nothing?" (*Warum ist überhaupt Seiendes und nicht vielmehr Nichts?*)[1] Although he did not provide a straightforward answer, Heidegger viewed the question as an essential part of the human experience and understanding of existence. Only by experiencing nothing and by questioning nonexistence can human beings understand their own finite being and the existence of objects and entities in the world.

Inspired by Heidegger, whose works he studied in a Nazi prison camp during the Second World War, the French philosopher Jean-Paul Sartre put his own stamp on the question of nothingness. In *L'être Et Le Néant* (Being and nothingness) he proposed that nothingness is intertwined with human consciousness, thus providing the space for freedom.[2] It is this freedom that allows individuals to define themselves and give meaning to their existence through their actions and choices.

୬

Just as per the philosophers, existence and nonexistence must be considered in conjunction: a hole can be understood only in connection to its border. It is a kind of absence that we understand only in relation to the presence of something else. The holes in Emmental cheese, for example, are defined

Gaping hole of a cistern

by the cheese surrounding them (please do not call the holey cheese "Swiss cheese": there are dozens of Swiss cheeses; only the Emmentaler has holes). A hole in the donut is referred to by the donut around it. Is a cheese hole the same as a donut hole? Is the Nothing in the cheese identical to the Nothing in the donut?

This raises again the thorny question of "what actually is nothing?" And can it exist? Well, if nothingness exists, then by definition it cannot be *no*thing because only *some*thing can exist. If nothing were to exist, then even the concept of "nothingness" cannot exist. But if there's nothing, there's a void. And a void, after all, is something. Using everyday language, we get entangled deeper and deeper into contradictions and paradoxes.

For mathematicians a world in which nothing exists cannot exist because at the very least mathematical truths exist necessarily. In arithmetic (number theory), one plus one will always be two, no matter what. So, at the very least, mathematical identities must exist. On the other hand, set theorists usually consider the null set "the empty set" in their theories. For example, whenever something is claimed about, say, all combinations of a set of elements, or when the set of all sets is considered, the null set must of course be included. Thus set theory implicitly accepts the existence of nothingness.

Physicists and astronomers, on the other hand, discuss nothingness not as an abstract or mathematical concept but as a scientific phenomenon. We return to Aristotle who maintained that nothingness cannot exist. Nature abhors voids—*horror vacui*—and matter, due to its density, will always seeks to fill any empty space. Furthermore, since everything must be located somewhere, and every physical location is defined by the thing that happens to be there, nothingness, having no location, cannot exist. Hence every place in the universe must be filled with something. As soon as a void begins to open up, it will immediately be filled with substance (i.e., matter).

Today's physicists agree with Aristotle's conclusions. It is impossible to achieve nothingness, i.e., a perfectly empty hole, because no device can remove every single atom or molecule from a space. And even if multiple vacuum pumps could expel all matter from a container, it would be impossible to shield the interior, the hole, from external radiation. Moreover, quantum theory predicts the temporary existence of "virtual particles" that pop in and out of existence, even in what is considered empty space.

Astronomers and cosmologists discuss the concept of "vacuum," empty space, and the void in nuanced ways. We know that the observable universe is expanding, and galaxies are moving away from one another into seemingly empty space. But even interstellar space, the space between stars and galaxies, which is the closest one can get to a perfect vacuum, is not truly empty. In addition to radiation, it always contains at least a few hydrogen atoms per cubic meter. And let's not even get started on dark matter, the mysterious unseen matter that supposedly makes up about 27 percent of the universe's mass (see chapter 12). Oh, and by the way, even a space that contains a perfect vacuum contains something, namely, the perfect vacuum.

Among scientists, the cosmologists are the ones who deal with the most baffling questions about nothingness. Was there nothing before the Big Bang? And will there again be nothing after the Big Crunch? Traditionally, cosmologists believe that the universe began as a singularity, a point of infinite density and temperature.

But what was before? Well, time did not exist before the Big Bang, so there was no before. And what was outside that singularity? Well, that's the definition of singularity: there is no outside. So there's no before and no outside . . . just nothing . . . but no nothingness . . . and we're back to where we started!

No wonder Bertrand Russell, the British philosopher and winner of the 1950 Nobel Prize for literature, preferred to take the easy way out: "the universe is just there, and that's all." No further explanation is needed. We are left, however, with the queasy feeling that we are largely ignorant about nothingness.

9

IS JUSTIFIED TRUE BELIEF KNOWLEDGE?

Gettier's Problem

In a discussion about the nature of knowledge and belief, Plato had Socrates ask the mathematician Theaetetus whether "true opinion accompanied by reason is knowledge?" Theaetetus agreed: "precisely," he said.[1] This led the American philosopher Edmund Gettier in 1963 to redefine Plato's characterization of knowledge succinctly in three words: necessary and sufficient conditions for knowledge are "justified true belief."[2]

Let us consider a proposition (*P*)—for example, "it rains outside"—and a subject named Quentin (*S*). Plato maintained, as reformulated by Gettier, that *S* has knowledge about *P* if, and only if, the following three requirements are satisfied:

(1) *P* is *true*
(2) *S believes* that *P* is true, and
(3) *S* is *justified* in believing that *P* is true.

The third requirement needs some context. According to the Greek philosophers, Quentin must have reason to believe, either by his five senses or through a rational and convincing proof, that *P* is true. In this example, (1) and (2) may be satisfied because it does in fact rain outside and Quentin believes it. But how did *S* come to believe it? Maybe he stood outside and got wet and looked up to see raindrops falling. In this case, he was justified in believing that it rains. If, however, he was in a windowless room and only heard the weather report that said that it was raining, then he was not justified in believing that it rains. The radio could have been tuned to a station of a different city and reported on that city's weather. In that case, Quentin did *not know* that it rained outside.

If (1) and (2) are satisfied, but not (3), then *S*'s belief, although true, is poorly justified; it is not knowledge but just a lucky guess. This is the realm of religious, philosophical, political, or ideological belief systems. They do not count as knowledge.

If (1) and (3) are satisfied, but not (2), then *S* is in denial; he should believe *P*, but in spite of the evidence, he doesn't. Conspiracy theorists—pushing denial of true *P*—have easy play with people like *S* who simply have no idea (aka no knowledge). Anti-vaxxers, climate-change deniers, and flat-earthers belong to this group.

If (2) and (3) are satisfied, but not (1), one does not have knowledge but a mistaken belief in a false statement. After all, one can only know things that are true; one cannot have knowledge about, for example, news that is fake. Again, conspiracy theorists—pushing a false *P*—have easy play with *S*. Mistaken beliefs are the realm of romance scammers who make naïve (ignorant) subjects believe their supposed proofs of affection. Believers in homeopathy, psychic powers, and alien abductions also fall for fake news and "alternative facts."

So according to Socrates, Theaetetus, and Plato, only if (1), (2), and (3) are satisfied simultaneously can *S* be said to have knowledge about the truth of *P*. But do (1), (2), and (3) suffice?

This is where Gettier begged to differ. He agreed that the three requirements are necessary in order to possess knowledge, but he claimed that they are not sufficient. In other words, even

Edmund Gettier

if the three requirements are satisfied, Quentin may, nevertheless, be ignorant of *P*.

Let me illustrate with an example: the time is 2 P.M. (14:00 in Europe). Tony looks at his watch and sees that it shows two o'clock. So the three requirements are satisfied: (1) it is 2 P.M.,

(2) Tony believes that it is 2 P.M., and (3) his belief is justified because the watch shows two o'clock. If Theaetetus lived today, he would say that Tony knows the time of day. But unbeknown to Tony, his watch stopped working the previous night at 2 A.M. So Gettier disagrees with Theaetetus: Tony does not know that it is 2 P.M.; it is just by coincidence that he believes it is.

Thus even a true belief does not count as knowledge if it has been gained by pure coincidence or by dumb luck. Knowledge requires that one has arrived at the belief in a justifiable manner, namely, as the result of rational reasoning or by personal experience or sensation. A lucky guess does not constitute knowledge.

Philosophers have grappled with the problem posed by Gettier for decades. Some have proposed to add some condition other than the justification clause to truth and belief as the constituent components of knowledge. For example, some thinkers also attribute knowledge to animals such as to one's pets. Hence replacing the justification clause with a reliability clause,

(3') S's belief that P is true was produced
by a reliable cognitive process,

would allow not only humans to be to endowed with knowledge but (nonhuman) animals as well because reliable cognitive processes also convey information to animals.

Or one could keep the three original requirements but add an anti-luck condition:

(4) S's belief is not due merely to luck.

፠

Belief refers to a state of mind in which one accepts something as true or rejects something as false, often without concrete evidence or proof. Knowledge, on the other hand, typically implies a justified and true understanding of something based on evidence, facts, and logical reasoning. They both interact and influence each other: beliefs can shape the interpretation and acceptance of knowledge, and knowledge can challenge or modify existing beliefs. Although both play a role when making decisions, knowledge provides a more robust and reliable foundation than unjustified belief for an understanding of the world around us.

10

HARD BUT EASY

P vs. *NP*

Apart from everything else they can do, computers speed up computations. The fastest computer today can perform a quintillion flops (10^{18} floating point operations per second). But even that is turtle speed when one tries to decipher an encrypted message. Obviously, it generally takes longer to compute something when the input is larger. How much longer? That is an important question in theoretical and applied computer science.

In computer science, an algorithm is considered fast if the time required to solve a problem grows like a polynomial with the size of the input. Let us denote the size of the input (say, the number of digits) by n, and—considering a simple polynomial—assume that the required time grows like n^2, the square of the input. So to solve a problem whose input is ten times as long as another, the computer will take a hundred times longer (10^2).

This is considered manageable. What is considered unmanageable is if the time grows *exponentially* with the size of the input, like 2^n. A tenfold increase in the input would increase the run time for such algorithms by more than a million (2^{10}).

The class of manageable problems that are solvable in polynomial time is called class P. For example, until a few decades ago, no polynomial-time algorithm existed that could verify whether an integer Z was prime or composite. Then, in 2004, three Indian computer scientists published the bombshell paper "PRIMES Is in P," in which they presented an algorithm to determine primality of a number; the algorithm's run time increased only like a polynomial as the digits of the investigated number increased.[1]

But even before 2004, mathematicians had a trick up their sleeve. Although it is hard to find the factors of a large integer Z, one can easily verify whether Z is a composite number if one of the suspected factors, say F, is given: simply divide Z by F, which can be done quickly; if the division is possible without leaving a remainder, then Z is composite (i.e., it is the product of F with another integer). Of course, that is nowhere near an answer to the original question (is Z prime?), but at least it is something. And best of all, the verification can be accomplished in polynomial time.

Let us take a step back: the class of problems where the correctness of a suggested solution can be verified in polynomial time is called the class *NP*. (*N*ondeterministic *P*olynomial time algorithms, but don't worry about the exact meaning of the term.) It stands to reason that if one can solve a problem in polynomial time, one can also verify a solution in polynomial time. Hence the class *P* is contained in class *NP* (in mathematical notation: $P \subset NP$). But here is the million-dollar question: Is the opposite also true? Can every problem whose solution can be quickly verified also be solved quickly? Must there exist fast algorithms for every problem whose solution is quickly verifiable?

By the way, there is also the class of so-called *NP*-hard problems: these are at least as hard as the hardest problems in *NP*. *NP*-hard problems may or may not belong to *NP*; those that do belong to *NP* (where potential solutions can be quickly verified) are called *NP*-complete.

᠁

As yet the answer to whether the class *P* of problems is identical to the class *NP* is not known. It is the notorious question of whether *P* = *NP*. And this truly is a million-dollar question: in the year 2000, the Clay Mathematics Institute promised a one-million-dollar prize to the person who proves or disproves *P* = *NP*, where the class *NP* consists of decision problems (yes/no problems) whose "yes" answers can be verified in polynomial time.

The implications involve far more than a million dollars. For example, if it should turn out that *P* = *NP*, then most cryptographic protocols that are in use today could be hacked by computers and would thus be rendered useless. We can visualize this as follows. To find a key to a padlock is difficult; it's like looking for a pin in a haystack. But if you happen to possess a key, it is very simple to see if it fits: just try it. Now, if it were as easy to find the key as it is to simply try a key, locks would be useless.

Similarly, cryptography depends on procedures that are simple to do but difficult to undo. In computer science such procedures are called one-way problems—easy to check but difficult to solve; in mathematics they are called hard functions—easy to compute but hard to invert. Take semiprimes, numbers that are the product of two prime numbers. The multiplication of two large prime numbers to produce a very large semiprime is easy. The factorization of a semiprime, often several hundreds of digits in length, into its two primes is difficult even with the fastest computers.

Most current cryptographical methods are based on hard functions. Crucially, these methods are based on the assumption that hard functions exist. However, if $P = NP$, then hard functions do not exist. Finding the appropriate cryptographic key would be just as easy as putting a random key into the keyhole and checking to see whether it fits.

Luckily, it seems that we need not worry . . . at least for the time being. Most computer scientists assume that $P \neq NP$, which is tantamount to their belief that hard functions do exist and that cryptographic procedures with large enough keys are therefore secure. But we must not allow ourselves to be lulled into safety. First of all, hard is hard, but not impossible. The fact that so far nobody has been able to develop a polynomial-time algorithm to factorize large numbers does not mean that nobody ever will. Furthermore, advances in computer software and hardware (think quantum computing) may soon allow vastly faster processing speeds such that even exponential run times can be tackled. But it would be nice to know if $P = NP$ . . . even if for nothing else than the million dollars.

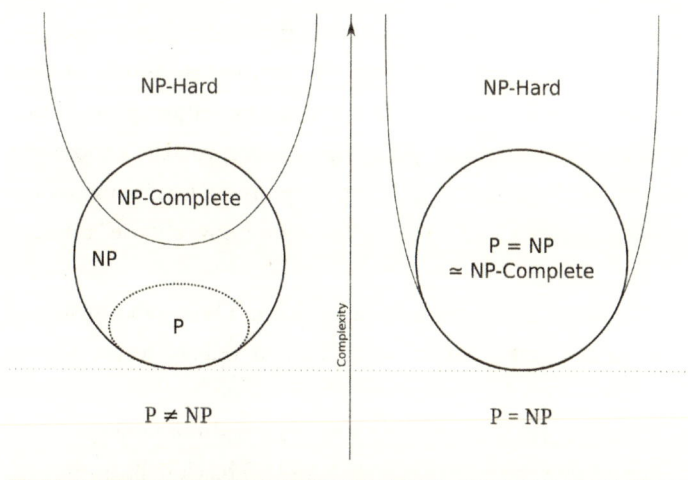

NP-Hard

NP-Complete

NP

P

$P \neq NP$

NP-Hard

$P = NP$
$\simeq$ NP-Complete

$P = NP$

Complexity

P vs. *NP*: Complexity classes

11

IGNORING ONE'S OWN ASSESSMENT

Ellsberg's Paradox

Suppose you have an urn with ninety balls. Thirty are red, the other sixty are either black or yellow. You don't know how many are black and how many are yellow; there may be 0 black and 60 yellow balls, 1 black and 59 yellow, or vice versa, or anything in between. If you guess correctly, you win a prize.

Would you rather bet on a red ball or a black ball?

There is no right answer, of course, and some people choose "black" and others "red." Let's assume that you choose red. (What follows is also true, *mutatis mutandis*, if you had chosen black.)

Before the ball is drawn from the urn, you are asked a second question: Would you rather bet on "red or yellow" or on "black or yellow"? Again, there is no right answer; you can choose one answer, or the other. But in experiments, at least some of the people who chose "red" in the first round now choose "black or yellow."

The latter group presents a paradox. Those who answered "red" to the first question and "black or yellow" to the second question contradict themselves.

Why? A priori, none of the answers is right or wrong.

Recall that there are between 0 and 60 black balls in the urn. So those who choose "red" in the first question—of which there are exactly 30—implicitly express their personal conviction that the number of black balls in the urn is less than 30. (It is completely irrelevant how many black or yellow balls the urn actually contains; we are only talking here about subjective beliefs.)

Thus anyone who believes that there are fewer than 30 black balls in the urn obviously believes that there are more than 30 yellow balls (since the black and yellow balls sum to 60). Together with the 30 red balls, the "red or yellow" balls add up to more than 60. And now, here's the kicker! Those who answered the first question with "red" but the second question with "black or yellow" violated their own conviction: as we shall see below, with the second answer they chose to ignore their own assessment as expressed in the first answer. This contradiction, expressed not necessarily by a majority of people but certainly by many, has gone down in history as the "Ellsberg Paradox."

⁊⁊

The name Daniel Ellsberg may sound familiar to many readers, but in a completely different context. Ellsberg (1931–2023) was behind publication of the infamous Pentagon Papers and narrowly escaped a decades-long prison sentence.

Born in Chicago in 1931, Ellsberg studied economics at Harvard University and then joined the United States Marine Corps. Back to Harvard, he earned a PhD in economics in 1962 with a thesis on risk, ambiguity, and decision-making.

In the mid-1960s, Ellsberg joined the RAND Corporation, a think tank in California, where he worked on the top secret study "U.S. Decision-Making in Vietnam, 1945–68." To his great horror, he discovered the government's deceptions and the ill-advised decisions, obscured by secrecy, under four presidents. He clandestinely photocopied the entire seven-thousand-page

Daniel Ellsberg in 2020

study and sent it to several newspapers. Subsequently, Ellsberg was charged with espionage and treason and faced a life sentence in prison. But because of horrendous misconduct by the prosecution, the judge struck down the charges.

A few years before he became infamous for the Pentagon Papers, Ellsberg had made a name for himself in a very different context. While still a fellow at Harvard, he published a paper showing that there were problems with the way some people deal with probabilities. "Risk, Ambiguity, and the Savage Axioms," published in 1961, is considered a landmark in decision-making theory.[1] The core of the paper was a test that Ellsberg had designed and tried out on his colleagues. It consisted of the two questions posed at the beginning of this article.

꒰꒱

Paradoxical choices arise because many people violate the so-called Independence of Irrelevant Alternatives (IIA) axiom of social choice theory. (In the spirit of this book, we might rename the axiom the *Ignorance* of Irrelevant Alternatives.) Formulated in the 1950s by Kenneth Arrow, the Nobel laureate in economics, the axiom states: When a decision is made between two alternatives, the addition of a third, irrelevant alternative should not change the choice. For example, suppose a person prefers pudding to ice cream as an after dinner dessert. Then, if fruit salad is suddenly added to the menu as an additional option, this person should not suddenly prefer ice cream. The addition of fruit salad as an additional choice is irrelevant and must be ignored.

Now recall that the first question was to choose between "red" and "black." The second question was to choose between "red or yellow" and "black or yellow." Rational people, regardless of what they think of the number of black balls, should compare "red or yellow" with "black or yellow" in exactly the same way that they compare "red" alone with "black" alone. The addition of "yellow" to the question should not influence their decision in any way. In Ellsberg's questionnaires, all subjects who initially chose "red" but reversed their choice once the "yellow" option was added violated the IIA axiom. Instead of ignoring the irrelevant alternative, they ignored their prior assessment.

Ellsberg, then thirty years old, unabashedly called to account his colleagues, often much older than he and well-known in professional circles, who participated in his experiment:

> There are those who do not violate the axioms, or say they won't, even in these situations (e.g., G. Debreu, R. Schlaiffer, P. Samuelson); such subjects tend to apply the axioms rather than their intuition, and when in doubt, to apply some form of the Principle of Insufficient Reason. Some violate the axioms cheerfully, even with gusto (J. Marschak, N. Dalkey); others sadly but persistently, having looked into their hearts, found conflicts with the axioms and decided, in Samuelson's phrase, to satisfy their preferences and let the axioms satisfy themselves. Still others (H. Raiffa) tend, intuitively, to violate the axioms but feel guilty about it and go back into further analysis.
>
> The important finding is that, after rethinking all their "offending" decisions in the light of the axioms, a number of people who are not only sophisticated but reasonable, decide that they wish to persist in their choices. This includes people who previously felt a "first-order commitment" to the axioms, many of them surprised and some dismayed to find that they wished, in these situations, to violate the [IIA axiom].[2]

&

Several mathematicians and statisticians had suggested that probabilities cannot be considered objective, and that people behave as if they make their own subjective judgments about uncertain events. Ellsberg begged to differ. His test showed that even subjective probabilities cannot be added up.

Are people who behave in this way irrational? Well, that depends on one's definition of rational. In any case, their behavior is inconsistent with the standard axioms of rationality. Ellsberg suggested that these people's decisions were motivated by ambiguity aversion: People seem to prefer taking risks in situations where they know the probabilities than when they ignore them. If they choose "black or yellow," then they know that there are 60 balls; if they had chosen "red or yellow," then there could have been anywhere between 30 and 90 balls.

12

IT'S THERE BUT CAN'T BE SEEN

Dark Matter

I n the 1930s, Fritz Zwicky (1898–1974), a Swiss astronomer working at the California Institute of Technology (Caltech), applied the so-called Virial Theorem in physics to the Coma Cluster, a large cluster of galaxies. The Virial Theorem (discovered in 1870 by the German physicist Rudolf Clausius) states that the total potential energy of a system, U (which is proportional to mass^2/distance), plus twice the system's kinetic energy T (which is proportional to mass × velocity^2) must add to zero:

$$U + 2T = 0.$$

Using the galaxies' redshift, Zwicky estimated the galaxies' velocities within the Coma Cluster and their velocities relative to Earth; measuring the brightness of certain types of stars and galaxies, he estimated the distances; and by comparing their "mass-to-light ratio" with similar types of galaxies whose mass had been estimated through independent means, he estimated the mass of the galaxies. With this, he had estimates for the three variables that he required.

After completing his computations, the astronomer was in for a surprise: nothing added up. Even after accounting for the many uncertainties and assumptions, the galaxies were moving too fast. If the mass that he had calculated were the only matter that provided the gravitational pull, the galaxies should have flown asunder. To keep everything in place, the mass had to be much greater than the mass of the visible matter.

Zwicky's conclusion was that since the gravitational force inferred from the galaxies' velocities was much stronger than what could be accounted for by the visible mass alone, there must be more, albeit invisible, mass hidden somewhere. Actually, the French mathematician and physicist Henri Poincaré had already coined a term for this puzzle: dark matter (*matière obscure* in French). Zwicky adopted the term (as *dunkle Materie* in German) when he published his findings in a Swiss scientific journal. He surmised that "the average density in the Coma system would have to be at least 400 times larger than that derived on the grounds of observations of luminous matter. If this would be confirmed we would get the surprising result that dark matter is present in much greater amount than luminous matter."[1]

The astrophysical community was aghast. Just like Isaac Newton's compatriots could not believe that there was an invisible force that pulled an apple down from the tree, it was hard for Zwicky's colleagues to imagine that the universe was filled with matter that cannot be seen.

The community's skepticism was understandable. After all, Zwicky's computations were just that: computations. Maybe they were wrong? (Indeed, they were, but only numerically because at that time only imprecise estimates were available of how quickly the universe was expanding.) What was required to convince the public of the existence of dark matter was observational proof. And that was about to come.

Kepler's Laws describe the dynamics of gravitational interactions between one or more bodies that revolve around a body in the center. Under the assumption that the body at the center is much heavier than the objects revolving around it, the famous Third Law describes how the orbital velocities of bodies change with the distance from the center. The law states that the speed at which, say, planets revolve around a sun, the so-called orbital period, is slower the further away the planets are from the sun. In our solar system, for example, about 99.86 percent of the total mass is concentrated in the sun. Therefore, Kepler's Third Law applies, and true enough, Venus, which is 109 million kilometers away from the sun, takes 225 days for one revolution, Earth is about 150 million kilometers away and requires a year, Mars, 245 million kilometers away, takes 687 days, and Saturn, 1.5 billion kilometers away, revolves around the sun once every thirty years.

Kepler's Third Law can be applied to any celestial system where bodies revolve around a heavy object in the center. Hence, in principle, the Third Law applies to moons orbiting a planet, planets orbiting a star, and also to stars orbiting the center of a galaxy: the further a star is from the galaxy's center, the slower it should move, according to Kepler's Third law.

But in the late 1960s, the astronomers Vera Rubin and W. Kent Ford discovered that this was not always so. In a paper in the *Astrophysical Journal*, they reported on the angular motion of stars within the Andromeda Galaxy. Instead of orbital periods increasing when stars are further away from the center, as Kepler's laws would suggest, they found that the outermost components of the galaxy traveled as fast as the ones close to the center. This could not be explained by the visible mass of the galaxy alone. Once again, the question arose: Is there more matter hidden somewhere?[2]

Rubin and colleagues examined dozens of galaxies, and their observations strengthened their suspicion that something other than the galaxies' visible mass determines the motion of the stars. They surmised that galaxies must be embedded in a "halo" of dark matter—material that does not emit light and extends beyond the optical galaxy. Their calculations showed that for Kepler's Third Law to hold galaxies must contain at least five to ten times as much dark matter as ordinary matter.

꒰꒱

Most physicists and astronomers have come to accept that dark matter exists, but it remains a mystery even today. It neither absorbs, nor reflects, nor emits visible light, nor any other electromagnetic radiation, such as radio waves, microwaves, infrared light, ultraviolet light, X-rays, or gamma rays. Were it not for the indirect effect of galactical gravitation, we would be totally ignorant of its existence.

The important question, apart from its very essence, is how much of this invisible dark matter there is in the universe. The consensus is that dark matter is much more abundant than visible matter. By today's state of knowledge, Zwicky's estimate of dark matter being four hundred times as abundant as visible matter was off by nearly two orders of magnitude. In order for the Virial Theorem to hold, based on modern more exact values of Hubble's Constant (which tells us how fast the universe expands), the current estimate is that ordinary visible matter—the stuff we can see and interact with—makes up only about 15 percent of all matter, and 85 percent of all matter, of which we are mostly ignorant, is dark.

(In memory of my friend Kurt Baumann, an amateur astronomer, of Switzerland and New York.)

13

UNSKILLED AND UNAWARE OF IT

The Dunning-Kruger Effect

Who has not had experiences with handy people who declared with confidence that they would be able to repair some defect in the house or the car only to make the problem worse? These handy persons do not cheat or lie; they are truly convinced that they can perform the task. The only problem is that they are ignorant of their lack of skill.

The problem, identified in 1999 by David Dunning and Justin Kruger, two social psychologists at Cornell University, is based on a cognitive bias, namely, the inability to assess one's own abilities. The phenomenon described in their paper, titled "Unskilled and Unaware of It: How Difficulties in Recognizing One's Own Incompetence Lead to Inflated Self-Assessments," has become known as the Dunning-Kruger effect. People with low ability or knowledge in a given intellectual or social domain often overestimate their competence, mistakenly assessing their abilities as much higher than they really are. The phenomenon—that some people are not only bad at what they do but also do not realize that they're bad—can be summarized succinctly, albeit unkindly, as "stupid people are too stupid to know that they are stupid" or, more politely, "unskilled people are too unskilled to know that they are unskilled."[1]

The effect was summed up laconically by David Dunning in an interview in 2019: "The first rule of the Dunning-Kruger club is you don't know you're a member of the Dunning-Kruger club." A mathematically absurd consequence of this cognitive bias is that if one asks a large enough group of random people about their skill level in any given field a majority usually believe that they score above average.

≈

The two psychologists investigated several groups of people, asking them to judge their sense of humor, their ability to reason logically, and their knowledge of grammar. To test their sense of humor, for example, subjects were asked to assess the quality of thirty jokes on a "funniness" scale; their ability to recognize what was funny was then compared to the ratings given by professional comedians. The test of logical reasoning used twenty questions taken from a preparation guide to the Law School Admissions Test, and to assess knowledge of grammar, students were asked to judge whether twenty sentences taken from the preparation guide for the National Teacher Examination were grammatically correct.

The results in all tests turned out to be as the authors thought. In the funniness test, the unhumorous subjects, i.e., those scoring in the bottom 12th percentile, estimated themselves to be way above average. Low-performing participants in the logical reasoning test also overestimated their ability: those who scored at the 12th percentile on average believed that their general logical reasoning ability fell at the 68th percentile. Grammatically challenged individuals who scored only in the 10th percentile estimated their abilities and performance on the grammar test to be in the 67th percentile.

The studies have been replicated in many different fields, such as driving skills, business, politics, medicine, and chess. Apparently, people with limited competence in a particular domain tend to overestimate their abilities, probably in the belief that the knowledge they do happen to possess is applicable elsewhere; or that they can pick up new skills more easily than others; or that they understand new information better than most. This lack of metacognition (the failure to assess one's own abilities or lack thereof) has several gloomy implications. People subject to the Dunning-Kruger effect may reach erroneous conclusions, make poor career choices, engage in dangerous sports, or expose themselves to social opprobrium, all the while believing that they are doing really well.

There's more: incompetent individuals lack the very expertise they need to recognize their lack of expertise. Hence not only do poor performers perform poorly on their tasks, but their inability to recognize their incompetence prevents them from trying to acquire the necessary competence. And without acknowledging that they are unskilled, they have no incentive to improve their skills.

There's still more: the inability to acknowledge one's own incompetence is coupled to an inability to recognize competence in others. After all, a prerequisite to judging a person's proficiency in a subject is to be competent in that very subject. So, if (a) one cannot judge oneself, and (b) one cannot judge others, then (c) one cannot compare oneself to others, and (d) one cannot learn to improve.

Under the subtitle "It Takes One to Know One," the authors concluded that "the same incompetence that leads [incompetent people] to make wrong choices also deprives them of the savvy necessary to recognize competence, be it their own or anyone else's. . . . One of the ways people gain insight into their own competence is by comparing themselves with others. We reasoned that if the incompetent cannot recognize competence in others, then they will be unable to make use of this social comparison opportunity."

※

There may be a silver lining. Under some circumstances ignorance may be bliss: optimism and overconfidence may spur people on toward their goals, helping them achieve even unrealistic ones. Perhaps meta-ignorance—not knowing about all the obstacles and complications along the path to one's goal—can empower the blissfully ignorant toward decisive action and enable bold steps forward.

14

THE HARD PROBLEM AND THE EASY PROBLEM

Understanding Consciousness

Throughout history the human mind and consciousness have been mysteries. Ancient thinkers such as the Greeks Plato and Aristotle and the Chinese Zhuangzi, theologians such as the Catholic priest Thomas Aquinas, the Islamic physician Ibn Sina (Avicenna), the Jewish rabbi Maimonides, the philosophers René Descartes, John Locke, and Immanuel Kant, the psychologist William James, and the psychoanalyst Sigmund Freud spent lifetimes puzzling about these problems.

Cogito, ergo sum (I think, therefore I am), Descartes's first principle, asserts that the act of thinking is irrefutable evidence of one's own existence. It suggests that when we think, doubt, or engage in any mental activity we are aware of ourselves as the thinking subject. This self-awareness is a central aspect of consciousness.

In modern times, it is mainly the neuroscientists and philosophers who wrestle with questions about brain activities and consciousness: how do we perceive, act, react, and interact with our environment? To gain an understanding of the various cognitive functions and abilities associated with consciousness—information processing, remembering, learning, focusing, problem-solving—scientists use methods and techniques such as fMRI, EEG, and PET scans to study the brain and its functioning. How do electrochemical processes influence brain activity, how do drugs affect attention, how do different training regimes improve skills? What is the role of the medial temporal lobes in forming new memories, the function of Broca's area for speech production, the purpose of Wernicke's area in the cerebral cortex for language comprehension, the purpose of the hippocampus in memory formation, and the role of the frontal and parietal lobes in attentional processes?

Let us take vision as an example. Vision involves the conversion of light into neural signals that the brain then interprets in order to construct our visual perception of the world. Researchers examine how light-sensitive cells in the retina encode visual information, how various visual areas in the brain process different visual attributes such as color, motion, and form, how we visually navigate the environment, and how we can understand the interaction between attention and vision. All of these problems and questions are tackled by scientists.

One model, namely, the Global Workspace Theory (GWT), postulates a virtual workspace in the brain where cognitive systems interact. The systems process information from sensory input and from memories in an unconscious manner, and then compete for access to the global workspace. Once access is granted, this information is experienced consciously and the workspace makes the information available globally by broadcasting it to motor systems etc. to coordinate the human's actions, decisions, problem-solving, and learning. GWT successfully models a number of characteristics of consciousness, although not consciousness itself.

⁂

Enter the philosophers. Attempts to understand how the mind works based on natural processes are the easy problems says the Australian thinker David Chalmers. They are easy (which is not to say simple or trivial) in the sense that they are deemed to be solvable, at least in principle, by using existing methodologies of neuroscience, psychology, physics, chemistry, and other fields. Scientists can employ empirical observations, experiments, and measurements to understand the mechanisms and functions of consciousness.

The hard problems consist of explaining the nature and existence of subjective experience (see chapter 30 on Qualia). How and why do physical processes in the brain give rise to the subjective experience of being? How can objective, third-person descriptions of brain processes explain the subjective, first-person experiences that are characteristic of consciousness? Why do we have qualitative personal experiences, and can they be reduced to physical processes?

We know more or less how physiological processes in the nervous system affect the brain, but we are still largely ignorant about how feelings arise in human beings. The view of a painting, the taste of a dish, or the sound of a melody may give rise in a person to feelings of happiness, sadness, elation, or depression. Such subjective experiences may not be felt, or may be felt differently, by other individuals even though they are subject to the exact same sensual stimuli.

Hence the hard problem of consciousness is to understand how brain processes give rise to subjective experiences. Why and how do electric currents in the brain, or chemical reactions that transmit signals between neurons, or hormones that modulate neuronal activity give rise to the subjective experiences of being cheerful, miserable, angry . . . ? What are the underlying mechanisms that transform these biochemical and bioelectrical processes into conscious feelings and perceptions of emotions? How do they produce the subjective, qualitative experience, the inner, personal aspect of consciousness that we all experience but find difficult to explain or measure objectively?

Integrated Information Theory (IIT), for example, addresses the hard problem of consciousness by measuring how much the internal connections in a system contribute to a unified and meaningful whole (the so-called intrinsic causal power of the system). It posits that consciousness arises in any system whose components integrate information. IIT quantifies the level and quality of consciousness using a measure called integrated information, which represents the amount of additional information generated by a complex of elements above and beyond the information generated by each of its parts alone.

IIT offers an explanatory framework for the existence and quality of subjective experiences, but its validation as a model of consciousness is still a subject of ongoing research and debate. One critique of IIT is based on the so-called Chinese Room Thought Experiment, devised by the American philosopher John Searle. A person who is able to manipulate Chinese symbols according to instructions she received, but without understanding the symbols' meaning, illustrates that integrated information alone may indeed simulate intelligent behavior but fail to explain the qualitative, first-person experience of being conscious. Hence, even if a system achieves a high level of integrated information, this does not guarantee genuine understanding or subjective experience. In this view, IIT risks equating the structural arrangement of information with actual conscious experience. In the fall of 2023, more than a hundred renowned scientists published an open letter claiming that IIT is untestable and therefore nothing but pseudoscience.

≈

Back in 1998, David Chalmers and Christof Koch, a well-known neurophysiologist, entered a bet. Koch and the biophysicist

Francis Crick (of Double-Helix fame) believed that consciousness was scientifically tractable. In particular, they thought that firing brain cells formed the neural underpinning of consciousness. Chalmers, however, was convinced that science alone would never be able to supply answers to the big question. The bet was

David Chalmers

Christof Koch

that within twenty-five years researchers would discover neural patterns underlying consciousness, which would have proved Koch's point of view.

In 2023, Koch admitted defeat—attempts of IIT notwithstanding. Undeterred, he is still convinced that consciousness will one day be understood scientifically. The only reason he lost the bet, he claims, is because nobody had been able to provide scientifically verifiable proof of consciousness mechanisms during the previous quarter century.

≥ё

So where do we stand today with respect to the hard question? What is consciousness, and why do we experience it? Is it a by-product of complex computation, or is there more to it? How is it that mental states, such as beliefs and desires, can cause physical states in the body? Why does seeing red feel like something? How should we study and understand the structure of our conscious experience?

Proposed answers to the hard question range from the sublime to the ridiculous: mystical traditions (religion), panpsychism (everything has some form of consciousness), nondual philosophies (mind and body, subject and object, do not form true dichotomies), monism (mind and body are two aspects of a more fundamental substance that is neither purely physical nor purely mental), psychological ether theory (consciousness existed before brains existed), deeper entity theory (subjects and experiences are entities beyond spacetime), and quantum theory (don't even ask).

Finally, some philosophers and scientists argue that the hard problem might be altogether unsolvable because of the very limitations of human cognition. This answer, the *non plus ultra* of scientific and philosophical understanding of consciousness, harkens back to the puzzling problems and vexing paradoxes of self-reference such as, for example, the Barber's Paradox: does a barber who shaves all men in the village who do not shave themselves, and only those, shave himself? Thus consciousness may forever remain a paradox.

15

UNCHARTED DEPTHS

Aqua Incognita

When the French science fiction writer Jules Verne penned *Twenty Thousand Leagues Under the Sea* in 1869, he envisaged the submarine *Nautilus*, piloted by Captain Nemo, venturing into the unknown depths of the oceans. It is a real adventure story because nothing was known at the time about the deep sea. Surprisingly, even today much of the deep sea is still unknown. Total darkness, enormous pressure, cold temperatures, and often unpredictable currents pose immense challenges; few explorers have ventured down. It is sometimes said that more is known about the surface of the moon than about the floor of the deep sea.

The deepest point on Earth, nearly eleven kilometers below the ocean's surface, is the Challenger Deep in the western Pacific Ocean. It was visited for the first time in 1960 by Jacques Piccard and Don Walsh, two oceanographers who traveled in a bathyscaphe, a self-propelled submersible for deep sea dives designed by Piccard's father, the physicist, inventor, and explorer Auguste Piccard. The bathyscaphe consists of an observation gondola, suspended from a hull that contains tanks that are flooded with water to make it sink, and tanks filled with gasoline (which is lighter than water) to make it surface. To withstand the immense pressure of more than one thousand kilograms per square centimeter, i.e., more than one thousand atmospheres, the gondola was constructed of twelve-centimeter-thick steel. To provide the scientists with air to breathe, an apparatus was installed that absorbs the carbon dioxide of exhaled breath and allows rebreathing of the unused oxygen.

અ.

It took five hours for the bathyscaphe to descend. When it landed, the explorers caught the first glimpse of the ocean floor. They perceived a mixture of algae first, and then, more important, had a revelation: there was life at that depth. Something like a flat fish looking like a flounder, about 30 centimeters by 15 centimeters, seemed to swim away from the bathyscaphe's lights. They were unable to see more because a cloud of stirred up sediment rose up from where the bathyscaphe had touched down and obscured their window. Modest as this achievement was in terms of scientific achievement, it at least established that living creatures could exist in total darkness at this enormous pressure. More a feat of technology than of science, the dive nevertheless went down in history as a testament to human ingenuity.

Since then several more unmanned as well as crewed visits to the depths of Challenger Deep have taken place. Using remotely operated vehicles, autonomous underwater vehicles, and manned submersibles, scientists tried to understand biodiversity and biogeochemical cycles, create detailed images of the rock layers, study the RNA of pico- and nano-plankton, and understand how deep sea ecosystems function. Sophisticated machinery scooped up sediment cores at the bottom of the sea, collected dozens of specimens of scavenging crustaceans, recorded videos and photographs of amphipods feeding on bait, taped sounds within the trench, collected water samples, and identified hydrothermal activity. Minerals and compounds not found elsewhere were discovered, as well as microorganisms that have unique metabolic pathways that allow them to survive in environments previously believed to be too hostile for any kind of life. Researchers have found surprising varieties of life forms that have developed adaptations to survive the extreme conditions of the deep sea, but no fish were ever observed; this throws the veracity of Piccard and Welsh's observation of the famous flatfish somewhat into doubt.

Nevertheless, much of the deep sea still remains, *terra*, or rather, *aqua incognita*. Although 70 percent of the surface of the Earth is covered by oceans, historically the priorities were first to explore the dry surface and then outer space. For now, the deep sea remains a treasure trove of largely undiscovered scientific wonders and mysteries waiting to be unraveled. Findings of future expeditions may yield insights into evolutionary processes, how creatures cope with the complexities of the environment, and allow a deeper understanding of the origins of life. After all, how could organisms adapt to an environment without light, immense pressure, and extreme temperatures?

But scientific efforts are costly. With no obvious commercial reasons, the significant financial investments and enormous technological resources required to explore the deep sea seemed unwarranted. This may change, however. The possible discovery of deposits of precious metals, rare earth elements, and other minerals are increasingly viewed as potential sources for future mining activities.

The search has begun for "polymetallic nodules," lumpy, golf-ball-size rocks, millions of years old and worth billions of dollars because they are rich in manganese, cobalt, nickel, and copper. Critical materials for almost all of today's electronics, clean-energy products, and advanced computer chips—from electric cars to advanced weapons systems—these metals are vital to the development of cutting-edge technologies and clean-energy solutions. The sea is believed to hold several times what land does of these rare metals. Also, sought after fuels, petroleum and natural gas, are believed to be stored under the sea floor.

In addition, bioprospecting, the search for new kinds of plants and unknown animal species from which medicinal drugs, biochemicals, and other commercially valuable materials may be obtained, some of which could have potential as pharmaceutical applications, will, sooner or later, also spark economic interest in the deep sea.

❧

Reducing our ignorance about the deep sea is also essential for today's most important means of communication, namely, the internet. After all, 99 percent of all intercontinental data traffic travels through undersea cables. Finding optimal routes to lay submarine cables requires detailed surveys of the largely still unmapped sea floor. Finally, tourism is also on the rise. The catastrophic implosion of the submersible *Titan* in June 2023 notwithstanding, businesses will certainly emerge to make deep sea tourism available to thrill-seeking masses in spite of the enormous risk of disaster.

But deep sea environments are very fragile and slow to recover from disturbances. Extractive activities, such as mining and drilling, raise significant environmental concerns, and infrastructure projects, such as tourism and the laying of cables, pose risks of significant environmental damage. Furthermore, we still know very little about the deep sea's role in climate regulation. Unrestrained human activities above the sea have led to catastrophic global warming, so maybe we had better leave the deep sea undisturbed.

II

CANNOT KNOW

The book's second part confronts the inherent limitations of human comprehension, acknowledging its boundaries. I discuss subjects that are unanswerable or unknowable, not always because of a lack of trying but because of the fundamental constraints of our perception and intellect. Some questions are unanswerable due to the limitations of nature itself, leaving us in perpetual ignorance.

16

ON LEARNED IGNORANCE

Cardinal Nicolas de Cusa

I t is commonplace among the faithful of all religions to ascribe the highest qualities to their deity. Muslims, for example, ascribe ninety-nine attributes to Allah: He is the most generous, the kindest, the most loving, the gentlest, the most praiseworthy, the most magnificent, and so on. For Jews, God is eternal, all-knowing, merciful, just, holy, loving, forgiving, compassionate, and majestic. To Christians He is omnipotent, omnipresent, omniscient, merciful, just, holy, loving, and gracious. The lists go on.

Of course, this presupposes that humans can envision His essence. But not all followers of any religion would admit that God's essence can be conceived. God is too complex and abstract for humans to contemplate, they believe; hence words alone cannot describe Him. An alternative manner of considering the supreme being, called "negative theology," is a school of thought that defines God's essence as what He is *not*. Instead of asserting, for example, that God is good, one would say that there is no evil in God; instead of declaring that God is omniscient, one would say that nothing escapes Him; and instead of stating that God is most benevolent, one would say that He is not malevolent.

Negative theology goes back at least to Plato who, in one of his dialogues, lets Socrates declare that "we know nothing about gods, neither about them nor about their names" (Platon, *Kratylos* 400d). In another dialogue, Timaios, a wealthy aristocrat and philosopher, asserts that "to discover the Maker and Father of this Universe were a task indeed; and having discovered Him, to declare Him unto all men were a thing impossible" (Plato, *Timaios* 28c).

Early religious thinkers thought similarly. A Christian theologian of the late fifth or early sixth century CE who wrote under the pseudonym Dionysius, and hence became known as Pseudo-Dionysius, asserted "that man is best united to God by realising that in knowing God he knows nothing about him" (*Myst. Theol.* I, as quoted by Thomas Aquinas). The foremost Jewish thinker of his time, the twelfth-century Rabbi Moses ben Maimon (aka Maimonides, or the Rambam), called on Jews to "recognize that He has no essential attributes" (*Guide for the Perplexed*, 1:50).

A century later, Thomas Aquinas, considered one of the Catholic Church's greatest theologians and philosophers, asserted that "it is because human intelligence is not equal to the divine essence that this same divine essence surpasses our intelligence and is unknown to us." He concluded that "man reaches the highest point of his knowledge about God when he knows that he knows him not, inasmuch as he knows that, which is God, transcends whatsoever he conceives of him" (*Quaestiones Disputatae de Potentia Dei*, Question VII, article V).

Of course, there were opposing views: natural theology bases any awareness of God on reason and science, and revealed theology relies on religious experience and epiphanies.

It was against this background that a fifteenth-century Catholic theologian and intellectual in Germany decided to put the discussion about the comprehension of God into another context, one that could not be dismissed or altered by philosophers who adhere to a different school of thought, namely, mathematics.

Nikolaus Krebs was born in Kues in Germany in 1401 and later took on the name Nicholas of Cusa, or Cusanus.[1] As a student, his wide-ranging interests spanned philosophy, law, sciences, metaphysics, astronomy, and mathematics, but he finally settled on ecclesiastical law. Although he was ordained a priest only in his thirties, he had been active in various church offices since his student days, visiting Paris, Constantinople, and Venice. Adept at church diplomacy, he managed to arbitrate disputes wherever he was called, thus gaining the respect of the ecclesiastical higher-ups. In 1448, the pope elevated him to the rank of Cardinal.

Unfortunately, as soon as he was invested with new powers, he showed himself to be hard and uncompromising. His idea of church reform was not one of modernization but, to the contrary, a return to stricter mores in monasteries and convents. His dictatorial bearing led to strife and even bloodshed. In particular, his rigorous edicts against the Jews were so harsh that they were deemed intolerable at the time even by council members, church colleagues, and the pope. Eventually he was recalled to the Vatican, and he died, frustrated and bitter, in 1464.

In his reflections about humans' understanding of God, Cusanus believed that it was arrogant to presume to be able to describe Him. Any positive statement about God would be inadequate, he wrote. By calling Him the most glorious, the greatest, the most magnificent, the most powerful, one simply does not do Him justice. In fact, such attributes diminish His status. Hence the correct way to approach God, Cusanus believed, was not by claiming to possess knowledge about Him but by realizing one's ignorance. Any attempt to understand God through concepts of human reason is hopeless.

In volume I of his three-volume work *De Docta Ignorantia* (On learned ignorance), Cusanus makes his point first by stating that humans can conceive of objects, beings, and things only in relative terms: entities can be known only by comparing them to what one already knows. And by "known" Cusanus means measured. Hence, to grasp something that is as yet unknown, one must compare it with, or measure it against, something that is already known.

According to his philosophy, humans can envisage only entities that can be measured; objects cannot even be conceived unless they can be ascribed by relative magnitudes, such as, for example, more or less. (*Nihil est nominabile, quo non possit maius aut minus dari . . .*) After all, if numerical values were abolished, Cusanus wrote, order, proportions, harmony, and all differences would vanish. In fact, beings and objects themselves would disappear if they cannot be assigned numerical values.

But—and this is Cusanus's second point—there are limits to quantitative comparisons. A case in point is Archimedes's procedure to compute the area of a circle by inscribing polygons into the circle and computing *their* areas. As the number of sides increases, the inscribed polygon approaches the circle. Thus the area of a circle (actually the number π) can be approximated by inscribing polygons with ever more sides and computing their areas. The crucial insight, however, is that the area (and π) can only be approximated, but never determined exactly.

To compute the area precisely, one would have to continually increase the number of the polygon's sides. But however many sides the polygon already possesses, an additional side can always be added. Hence there is no end to the procedure: infinity can never be reached, it properly belongs to God alone because God is "the infinite, absolutely maximum power."

Therefore, just like we remain ignorant of a circle's true area, we remain ignorant of the true essence of God. This is *learned ignorance*.

17

WHERE AND HOW FAST?

The Uncertainty Principle

D r. Quentin Mechanos, a nuclear physicist, is caught speeding. During the hearing at traffic court, he challenges the policeman: "Where exactly was I caught speeding and what speed was I going?" The officer looks at his notes and answers precisely: "We caught you on the Interstate, at 7.5 kilometers after the intersection, and you were traveling at a velocity of exactly 115 kilometers an hour." Dr. Mechanos turns to the judge: "Obviously that is not correct. The so-called Uncertainty Principle, one of nature's fundamental laws, says that one cannot determine precise position and velocity simultaneously." The judge, an avid reader of science books, agrees and dismisses the case.

It was this German physicist Werner Heisenberg (1901–1976) who formulated the Uncertainty Principle of quantum mechanics. The principle states that in the world of the very, very small (for subatomic particles, that is) pairs of so-called conjugate variables are linked in such a way that only one of them can be determined with high precision. The more accurately the one is measured, the fuzzier the other must be. One necessarily remains ignorant to a certain degree about the precise values of conjugate variables.[1]

Imagine a billiards table in a dark room; a ball rolls on the table and you want to determine its position and its velocity. You can hold out your hand and try to touch the ball. At the very moment that you feel the ball touching your hand, you know its position. But simply by touching it, you have slowed it down and changed its direction. In the same manner, by shooting photons toward an electron to "see" its position, the electron's velocity will be affected. The photons transfer some of their energy to the electron, thus changing its velocity. Hence it is impossible to determine a subatomic particle's position and velocity at the same time.

This phenomenon is known as the "observer effect." Heisenberg derived his Uncertainty Principle not from experimental observations but as a mathematical consequence of what happens when conjugate variables—or rather the operations to measure these variables—do not commute. What does it mean for operations not to commute?

Let us consider a book lying flat on the table, facing you. Operation A turns the book ninety degrees anticlockwise around the z-axis. The book still lies flat on the table, but now the spine faces you. Operation B rotates the book ninety degrees around the x-axis. The book now stands sideways, facing you, with the spine along the table. Now invert the two operations. Operation B rotates a book around the x-axis: the book is now upright, with the spine on the left. Operation A rotates around the z-axis: the book now faces to the right, with the spine facing you. Same operations but different results: $AB \neq BA \neq 0$; operations A and B do not commute.

An example of such noncommuting conjugate variables in the quantum world are position and velocity of subatomic objects. If the object's position is determined to within only a small error, the error in the velocity will be large, and vice versa.

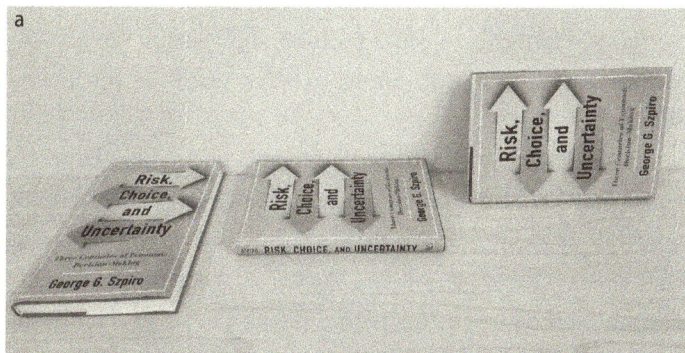

Noncommuting operations

The reason for that situation is that quantum objects can be described both as particles and as waves (neither of which can fully describe their behavior). When considering the object as a particle, its position can be determined with only a slight error by an operation, say C. When considering it as a wave, one can determine its velocity fairly precisely by operation D. But, as

Heisenberg showed mathematically, operations C and D do not commute. Hence the object's position and velocity can actually only be determined with errors, in other words, as probability distributions.

Heisenberg derived a specific formula about how large the errors must be at a minimum. He did not consider velocity by itself, but momentum (p), which is the object's mass (m), multiplied by its velocity (v): $p = mv$. We consider the mass to remain constant and denote errors with a Greek Delta (Δ). Hence the errors in position is Δx, the error in momentum is Δp. Heisenberg formulated the uncertainty relation thus: since the measurements do not commute, the product of the errors in the measurements of location and momentum must be greater than or equal to a minimum value, namely, $h/4\pi$, where h denotes the Planck-constant named after the German physicist Max Planck (1858–1947). Hence,

$$\Delta x \Delta p \geq h/4\pi,$$

or, since we considered mass to remain constant,

$$\Delta x \Delta v \geq h/4\pi m,$$

As befits nearly everything in quantum mechanics, Planck's constant has a teeny, weeny value:

$$h = 6.626\ldots \times 10^{-34} m^2 \text{kg/s}$$

&

So was the traffic judge right in dismissing the case, or did Dr. Mechanos pull a fast one? Let's go through the numbers.

We assume that the car has a mass of 2,000 kilograms and we also assume that the police determined the location of the car at the moment when its velocity was measured to within 100 meters (i.e., $\Delta x = 100$ m). Therefore,

$$\Delta v \geq h/(4\pi \cdot m \cdot \Delta x) = 6.626 \ldots \times 10^{-34} /$$
$$(4\pi \cdot 2000 \cdot 100) \; m^2 kg/s/kg/m$$

$$\cong 0.00000000000000000000000000000000000000000263 \ldots$$
$$km/hour$$

Hence, the police measured the speed at which Dr. Mechanos's car was traveling to within about 10^{-40} km/hour. The honorable judge should have read his science books a bit more carefully.

18

HALF ZEROS, HALF ONES

Random Numbers

When coins are tossed to kick off a football team, they fall half the times heads, half the times tails (or as mathematicians like to say, zero or one). In games with dice, the numbers between one and six appear, each with probabilities of one-sixth. And in a casino, the roulette ball falls on any number between zero and thirty-six about 2.7 percent of the time. What these tosses have in common is that the numbers thrown are random; one is utterly ignorant about the sequence of throws. Gamblers who believe that after several throws on tails a toss of heads must necessarily follow go bankrupt sooner or later.

Series of numbers that meet three criteria—all numbers thrown (or drawn) are equally probable, independent of the previous ones, and unpredictable—are called random, and they have great importance in many areas of daily life. (These three criteria define uniformly distributed random numbers. In other distributions, the numbers are not equally probable, such as in so-called normal distribution.) Random numbers are required in economics, medical research, science, mathematics, and computer science. For opinion polls, for example, candidates are selected by means of random numbers; in medical research, test subjects are randomly assigned to different groups; and in drafting recruits, the names are drawn by lot.

Random numbers are extremely important for the simulations of probabilistic processes. They indicate which of several alternatives may occur in a scenario. Such efforts began with the Manhattan Project in the Second World War, when Americans scientists developed the first atomic bombs. Nuclear explosions cannot be tested experimentally, so one had to make do with simulations. But not only dangerous phenomena are routinely simulated. In complex systems, the interaction of external and internal influences is often far too intricate for the effects to be calculated using the laws of probability. Aircraft manufacturers, for example, simulate the behavior of aircraft under different weather conditions and pilot reactions. Companies use simulations to run through scenarios to find out how procurement costs, collective bargaining, and strikes affect profits. Economists simulate how economic decisions influence each other and affect inflation and unemployment.

In computer science, algorithms can be accelerated tremendously with the help of random numbers. For example, in the 1950s, the so-called Bubblesort was the algorithm of choice to order a list of millions of names alphabetically. When Quicksort was launched in 1961, an algorithm that makes use of random numbers, the running time could be accelerated by several orders of magnitude. In addition to prime numbers, cryptography uses, in addition to prime numbers, simple numbers that are chosen at random. And with random simulations, even the irrational number π can be approximated by computing the proportion of random dots that fall within a circle.

In the coordinate system, a square with a side length 2.0 is defined around the zero point. The area of the square is 4.0. A circle of radius $r = 1.0$ is drawn inside this square; its area is

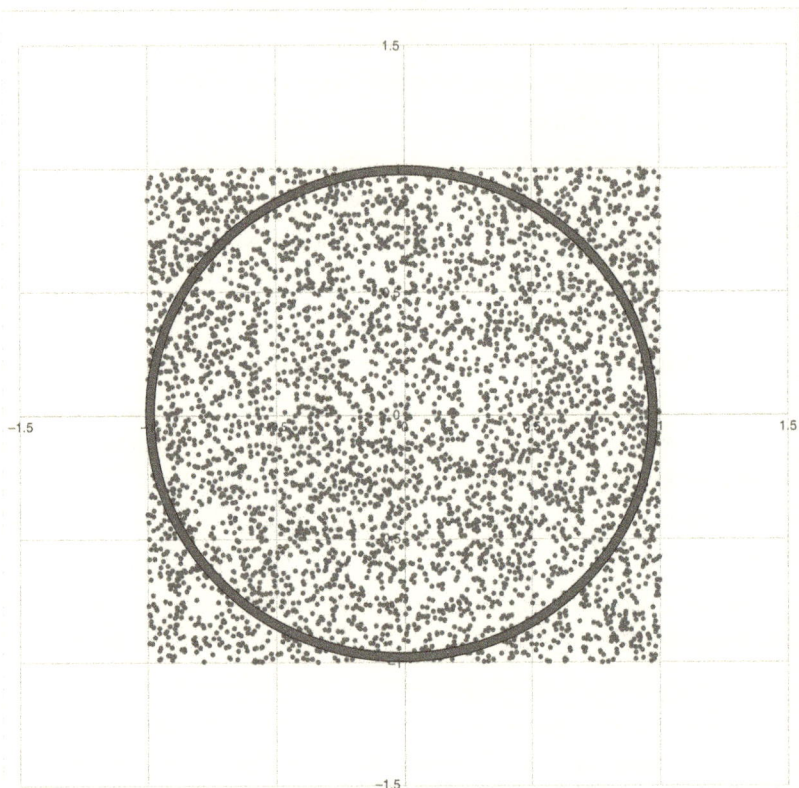

© George Szpiro

$\pi \times r^2$, which is equal to 3.141. . . . Now, if ten thousand random numbers between –1 and +1 correspond to the x and y coordinates of five thousand random points inside the square, approximately 78.539 . . .% of the points fall inside the circle. This proportion (78,539 . . .% of 4.0) equals about 3.1415 . . . which approaches π.

For many applications millions of random numbers are required and one immediate thought might be that they should be generated by computers. Unfortunately, this is not possible. Computers are deterministic systems; they must be told exactly what to do. Instructions such as "pick heads or tails" or "select a number between zero and nine" cannot be followed by a computer. The best that computers can do is to produce *pseudorandom numbers*. These numbers look as if they were randomly generated, but in fact they were produced by deterministic, albeit complex, algorithms. Since the days of the Manhattan Project, mathematicians and computer scientists have been striving to find ever better algorithms to generate numbers that meet the three criteria previously mentioned. The important objective is that one must always be ignorant of what the next number will be.

Unfortunately, what is common to all such computer algorithms is that the generated numbers are in some way related to the preceding numbers in the series. This means that one of the criteria—namely, independence—is always violated. To mitigate the effect of this violation, to some degree, one chooses the first number of the series (the "seed") as randomly as possible—such as the millisecond at which one clicks "Enter" on the keyboard— and then makes the relationship of further numbers with their predecessors as obscure as possible. This is accomplished with the help of *one-way functions* that are easy to compute (such as multiplying two numbers) but hard to reverse (such as factorizing a number). For a human being, it is then impossible to fathom how the members of a pseudorandom sequence follow one another.

Purists, however, are convinced that even pseudorandom numbers are inadequate. For example, in 1968, after several years of use, IBM's RANDU generator turned out to be flawed: projected into a three-dimensional space, the numbers are not random at all, but fall on several two-dimensional planes.

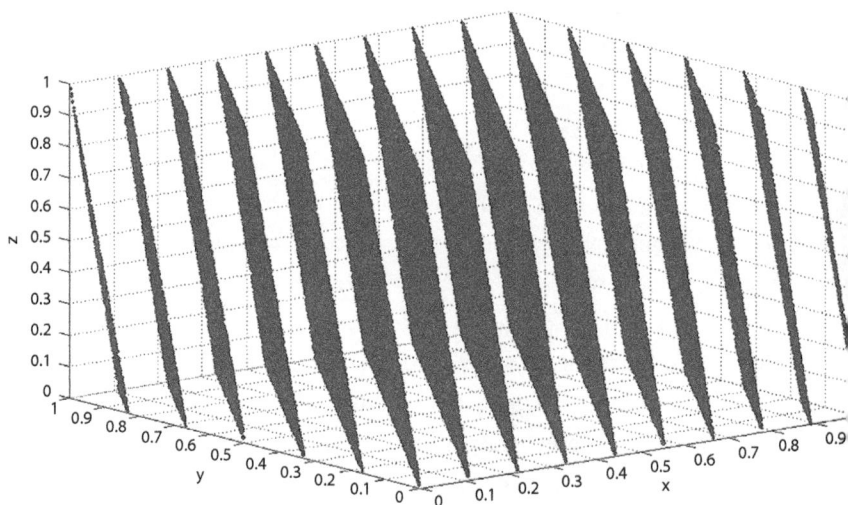

100,000 three-dimensional points created by RANDU. Image by author.

And even after a generator has been tested with all available tests, it might one day turn out that a specific pseudorandom number series fails in a very specific simulation. Such a mishap went down in scientific history as the Ferrenberg Affair. In 1992, the American Alan Ferrenberg simulated a model that physicists use to describe the behavior of ferromagnets. The results he obtained were completely incorrect. As it turned out, the random number generator that he used had subtle correlations that caused the random numbers to have a certain bias.[1]

That's why researchers often make do with numbers generated naturally, such as atmospheric noise, lava lamps, weather conditions, stock market prices, or human activity (the movements of a computer mouse or time delays in typing). Computer scientists have also shown how several time series, even imperfect ones, can be mixed to extract as much randomness as possible.

But even these methods are inadequate. After all, meteorologists and investment advisers often forecast weather conditions and stock prices with some success. And even tosses of coins and dice are not entirely random. Because they obey Newton's equations of motion, they are predictable, at least in principle: if initial conditions and the forces acting on them are known precisely, the outcome of tosses could theoretically be predicted. That the balls rolling on roulette tables do not quite obey the laws of probability has been known for a long time.

This leaves only one method that most scientists believe generates truly unbiased random numbers: quantum mechanics. The Swiss company ID Quantique in Geneva, for example, has developed a generator that makes use of quantum effects. Photons are fired at a semitransparent mirror placed at a 45-degree angle to the firing direction; half of the photons fly through the mirror; the other half are deflected. Photon counters register the bit 0 in one case and the bit 1 in the other. According to the laws of quantum mechanics, it cannot be predicted which case will occur, so the sequence of zeros and ones is completely random. But this is not completely acceptable either, because the unbiasedness of the sequence of bits depends on the apparatus. Among other things, to generate true random bits the mirror would have to be placed at an angle of exactly 45.00000000 . . . degrees, which is not feasible.

But can a sequence of purely random numbers exist? Paradoxically, we will never know the answer: if a series of numbers is indeed purely random, we could not recognize it as such because, if we did recognize it, it would not be random.

19

SATISFICE, DON'T OPTIMIZE

Bounded Rationality

Accordingto economic theory, decision-makers (that is people such as you and I and everybody else) should always strive for optimal decisions: maximize profits, minimize costs, take the shortest route, buy the best house, hire the brightest assistant, marry the ideal partner, and so on. This assumes, first of all, that one has gathered and processed all relevant information. And it often involves sifting through myriad alternatives; it also requires the ability to perform complex calculations. All of this must be done within a reasonably short time, and some decisions must be made on the spot.

In the search for the optimal spouse, one needs to date a multitude of potential partners; to find the most capable assistant, one must interview dozens of candidates; to buy the best house, one must gather reams of information; to decide on an insurance policy, one must perform complex probability calculations. Even everyday decisions such as what detergent to buy necessitates the comparison of prices and the qualities of competing products; to choose whether to take a bus, a taxi, or just walk requires weighing the tradeoffs between cost, time, and effort; to decide whether or not to take an umbrella one must evaluate the appearance of the sky and estimate the probability of rain based on past history.

All of these activities require complex operations. Data may not be available, and access to information may be limited. And even if all the necessary data were available, the need to process them often exceeds the computational abilities of humans. As a result, decision-makers often ignore much data and many numbers, and have recourse to shortcuts instead.

Let's say the profits of a corporation grew by 10 percent during the first year, by 20 percent during the second, and by 2 percent during the third year. What's the compound growth over two years? The correct result requires multiplying the individual growth factors, deducting one, and multiplying by one hundred:

$$((1.1 \times 1.2 \times 1.02) - 1) \times 100 = 34.64\%.$$

Hence the compound growth is 34.64 percent. However, most people find it easier to add than to multiply, so they use a rule of thumb: simply add 10 percent to 20 percent to 2 percent to obtain the approximate result, namely, 32 percent:

$$10\% + 20\% + 2\% = 32\%.$$

The error (2.64 percent in this example) is so small that it can often be safely ignored. The conclusion is that in many cases a rule of thumb, such as adding instead of multiplying, may provide a satisfactory solution at a lower cost of computational processing. Rules of thumb and other shortcuts are called *heuristics*. They are based on common sense and educated guesses that, in turn, are based on learning, experience, and feedback.

The use of heuristics must not be confused with being irrational. To the contrary, when one includes the efforts required to collect data, process information, consider all facts, and compute the probabilities in the decision process, the use of heuristics is perfectly rational. Given the limited abilities of humans, it is perfectly rational to ignore certain facts and numbers in exchange for simplicity and ease of computation. Herbert Simon, the psychologist/computer scientist/sociologist/political scientist who won the Nobel Prize in economics in 1978, named the strategy of trading off the search for the optimum against the cost of deliberation *bounded rationality*.

Another term that Simon introduced was the verb "to satisfice," which indicates that decision-makers do not—or should

Herbert Simon

not—always optimize but, instead, should set themselves a target that will satisfy their objective and stop the search when that target has been reached.[1]

What are the heuristics that people use when making decisions? The Israeli psychologists Amos Tversky and Daniel Kahneman spent years of pathbreaking research assessing people's decision-making habits. (Kahneman was awarded the Nobel Prize in economics in 2002. Tversky had passed away at age 59 in 1996.) They identified several shortcuts and rules of thumb, together with the biases that result as a consequence. People generally use the information at hand, be it pertinent or immaterial; they believe evidence if it confirms their prejudices; they exaggerate the importance of vivid over bland evidence; they ignore new information; they assign high probability to random events fresh in their minds; they rely with overconfidence on their own judgment; they think they are able to discern patterns in what is actually random data; they take into account sunk costs; and they fail to ignore irrelevant alternatives (see chapters 38 and 44.)

One of their best-known psychological experiments concerned the notion of framing. Depending on how a decision problem is worded (or framed), the answers that are provided can be diametrically opposed to one another. Let's say that an outbreak of a disease is expected to kill six hundred people. Fortunately, there are treatment options, vaccinations A and B, but they have their own risks. People are asked which vaccination they would prefer:

Vaccination A: two hundred people will be saved.
Vaccination B: two-third probability that nobody will be
            saved, and one-third probability that six
            hundred people will be saved.

Many people prefer A to B.

Now, the same two vaccines are available, this time we'll call them A' and B' but the question is framed differently:

Vaccination A': four hundred people will die.
Vaccination B': one-third probability that nobody will die, and
          two-third probability that six hundred people
          will die.

This time many people prefer B' to A'. The people who were surveyed ignored that A and A' are identical, and so are B and B'. The conclusion is that decisions can be influenced by the way the question is framed.

In 1979, Kahneman and Tversky published a pathbreaking paper in the prestigious journal *Econometrica*, titled "Prospect Theory: An Analysis of Decision Under Risk."[2] It spawned a full-fledged subfield in behavioral economics. I won't describe all of the details except to say that people tend to round low probabilities down to 0 percent, and round high probabilities up to 100 percent. The small errors are routinely ignored.

20

DON'T EVEN ASK

Meno's Paradox

What's a hexakaidecahedron? Is it an insect with more than a dozen legs? Is it an incantation used in witchcraft? If I told you, it is an outer planet of the solar system, would you believe me? If you were told it is a sixteen-faced polyhedron, would that be correct?[1]

More generally, is there a point in asking the question? Or of querying Google, Wikipedia, Stack Exchange, or ChatGPT?

❧

If you opted for the polyhedron, you were, in fact, correct. But how did you know it was the right answer? Maybe you studied advanced geometry in high school. Or you read my book on *Kepler's Conjecture* in which the hexakaidecahedron makes an appearance.[2] Maybe you heard the term before and knew what it was. In all of these cases, you did not really need to inquire about the term's meaning because you already knew the answer.

But what if you did not know beforehand what a hexakaidecahedron is, and someone simply told you "It is a sixteen-faced polyhedron"? How would you know whether this is the correct answer? What if someone had told you "It is a sixteen-sided polygon"? How would you know that this is an incorrect answer? You would not. Hence, if you don't know the answer beforehand, you would not be able to recognize a response as correct, even if it hit you over the head.

The upshot of all this is that if you already know the answer to a question, there's no need to search for it. Hence you cannot learn anything by asking. And if you don't know the answer, you would have no way of recognizing a correct answer even if it were given to you. Therefore, there's no point in asking questions.

ह

As reported by Plato, the question of whether there is a point in even asking a question arose in a debate between Socrates and a young general by the name of Meno. Discussing what is "virtue," Socrates challenges Meno to give a definition of the term. After several unsuccessful attempts, Meno is about to give up. But suddenly he has an inspiration. In an attempt to upstage Socrates, he challenges the philosopher in return. How would he, Socrates, recognize the correct answer if he did not already know what virtue was?

According to the general, a thinker who searches for answers finds himself in a dilemma: he cannot search for what he knows—after all, if he knows it, then there is no need to search for such a thing. And he cannot search for what he doesn't know—after all, if he doesn't know it, he does not even know what he's searching for. Well-pleased with his brainwave, Meno gloats: "Well, doesn't this argument seem to be finely stated, Socrates?"

But Socrates's rejoinder follows immediately: "Not to me!"

Of course, Socrates does not agree. The hallmark of his method of inquiry, the so-called dialectical method, is to elicit truth by dialogue until the interlocutor "gets it." To illustrate his method, Socrates draws some geometrical figures in the ground and then leads one of Meno's ignorant slaves through a series of steps until this untaught man realizes a geometrical truth: the area of a square can be doubled by constructing a new square whose side is the diagonal of the original square.

The slave did not know the answer beforehand, nor did he know what to ask. Nevertheless, when the correct answer becomes obvious in the course of the to-and-fro between Socrates and the slave, it just hits him. Hence knowledge can be acquired through guided reasoning, and truth can be made explicit through dialectical questioning.

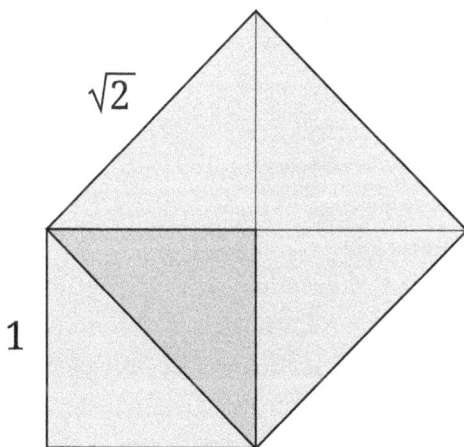

By drawing a diagonal through the original square, a new square can be constructed using the diagonal as its side; the area of this new square will be twice the area of the original square.

But there is a missing link: this argument assumes that the slave is not totally ignorant. He does not know enough to find the answer on his own, but he does know enough to recognize it as correct when it is presented to him. How does Socrates explain that?

Well, says Socrates, the slave, like everyone else, possesses prenatal knowledge, an understanding that is imparted to the immortal soul of every human being before he or she is even born. In rational inquiry, a human being can call on this knowledge when needed through the process of recollection. Thus one comes to know what one did not know previously.

≥≥

This explanation does not sound very convincing. An appeal to the immortal soul and to prenatal knowledge seems a bit far-fetched. Indeed, one may be excused for thinking that it is a cop-out. After all, the slave did not draw on the alleged prenatal knowledge all by himself to recognize a truth. By guiding the slave to the correct answer with leading questions, it was the philosopher who played the role of what he termed "prenatal knowledge" in order to elicit the knowledge that was supposedly latent within the slave.

In the ensuing discussion, even Socrates admits that he is not quite sure about his theory of prenatal knowledge and recollection. But there is one thing about which he is certain: he vehemently emphasizes that "we shall be better, braver, and more active men if we believe it right to look for what we don't know."

≥≥

Socrates's own doubts notwithstanding, his theory of prenatal knowledge was revived with Noam Chomsky's theory of language acquisition. How is it, the MIT-linguist asked, that children, supposedly born as blank slates, learn to speak? And how is it that as they grow they learn to express things of which they have no empirical knowledge? Chomsky holds the view—not universally accepted, however—that the human faculty of language is innate, hardwired in the brain, a view that hearkens back to Socrates's notion of prenatal knowledge.

21

ENGAGING IN RISKY BEHAVIOR

Moral Hazard

The parties to an insurance contract set up an agreement that generally covers all conceivable consequences. A well-thought-out contract tries to foresee all eventualities, but as in all business ventures some degree of risk remains. Indeed, the risks of, and the uncertainty about, future events are the reasons insurance contracts are concluded.

However, it is often the case that one party is aware of a fact that the other party does not know about and that is therefore not included in the written agreement. Or one party may know that the probability of an adverse event is greater than the other side thought it to be. For example, the policyholder may have a predisposition for a particular disease, a fact she did not disclose to the insurer. On the other hand, the insurance company possesses statistical information that may indicate that a certain risk is actually much lower than the purchaser believes to be the case. Just think of advertisements that entice people to purchase additional insurance against risks that are all but negligible.

Even more concerning is the tendency of insured individuals to engage in risky behavior; or to fail to institute preventive measures; or in case of a mishap to avoid limiting the damages. The fact that an insurance contract has been signed may tempt insured parties to act in an irresponsible manner, thus increasing the probability of an adverse occurrence and raising the damages beyond what the insurer had considered. Because insured individuals are shielded from the fallout, they do not care about the harms their behavior causes. This does not necessarily constitute fraud; it is simply human behavior. Nevertheless, such behavior may be considered wrong, and for this reason a special term has been coined: whenever information is asymmetric—one party to an agreement knows something that the other side ignores—then one speaks of "moral hazard."

Hazard means danger, risk, or threat. But why moral?

Originally, the term did not have the meaning of moral, as in ethical. When the Swiss mathematician Daniel Bernoulli (1700–1782) investigated how people behave while gambling, he differentiated between the purely mathematical aspect of money and the enjoyment that such money provides to the gambler. He named the latter the "moral fortune" to distinguish it from the physical fortune. Today one speaks of the "utility of money" (or of wealth) to specify how useful it is to the person who possesses it. The term *moral* distinguishes between the objective value of money and the subjective utility of money.

Similarly, the terms *real* and *moral hazard* express the imbalance between objective (statistically determined) risk and subjective (personally perceived) risk. Under an insurance contract, the insured party has little incentive to avoid risks, or take responsible action, or limit damages: it is protected against the fallout of adverse events because the other party will incur the cost. Hence the true extent of the risk the insurance company assumes is higher than statistical data would suggest. Moral hazard arises in situations when information is asymmetric.

Moral hazard is to be distinguished from another broadly related concept that also indicates asymmetric information, namely, "adverse selection." The latter refers to the fact that individuals who know that they represent a higher risk than the general population (for example, smokers) have a larger incentive to buy insurance (for example, against lung disease). The unsuspecting, ignorant underwriter is then stuck with a pool of clients who represent a higher risk than the average individual. The difference between the two notions is that adverse selection usually occurs before an agreement is signed, whereas moral hazard arises after the terms of the transaction have been agreed.

෨

Property insurance is the classic example of moral hazard lurking at every corner. When one's house and its contents are covered by an insurance policy, there is not much reason to lay out additional money for preventive measures. Homeowners could, of course, install smoke detectors, connect a burglar alarm, put bars in front of windows, and add a security lock on the front door. But why bother? Moreover, if an adverse event occurs, policyholders have a patently illegal tendency to inflate the damages that they allegedly suffered.

A less obvious area in which moral hazard looms is life insurance. Although suicides are excluded from insurance contracts (at least for a certain time period after the beginning of the policy), the insured may—unbeknown to the underwriter—pursue extreme sports such as parachuting, bungee jumping, deep sea diving, and rock climbing. Savvy insurance companies demand extra premiums to cover against accidents that can arise from high-risk activities.

Workers' compensation is another area where moral hazard appears. It is fairly easy for employees to fake illnesses or temporary disabilities, to exaggerate hurts and aches, or to prolong

Bungee jumping is usually not covered in insurance contracts.

their time of recovery. To track down cheaters, employers who are of course ignorant of their workers' true state of health engage trusted doctors to examine sick employees. If serious fraud is suspected, detectives are hired to track suspected cheaters, and if the suspicions turn out to be true, lawsuits can be filed in court.

There is more: company cars and office technology are handled with much less care if they are the property of the employer than if the employees themselves owned them. And speaking of cars, people have the tendency to drive more recklessly when the vehicle is insured.

꩜

Insurance is not the only industry in which moral hazard occurs. The finance industry is replete with occasions for moral hazard. Investors know only what managers choose to reveal about a corporation's financial health. In their ignorance, they must interpret publicly available clues and cues as best they can. To counteract the asymmetry in information that is available to managers on one hand, and investors on the other hand, the Securities Exchange Commission enforces regulations against insider trading through criminal prosecution (see chapter 34).

But then again, shareholders themselves are protected by the laws of limited liability. In case of a bankruptcy, the corporation's ignorant creditors are left to pick up the pieces (see chapter 34 on the Corporate Veil). Another risk of financial moral hazard occurs, for example, when mortgage lenders pool the loans, bundling good risks with bad ones, and then go on to sell pieces of this mortgage-package to investors, thus passing the risk of default on to others who are ignorant of the true risks. This was the reason for the subprime meltdown and financial crisis of 2008.

꩜

How can insurers, employers, or investors avoid or at least diminish moral hazard? Exclusion of certain risks, such as suicide or high-risk sports, is one solution to the problem. Property insurers insist that the insured party install protective equipment and take damage reducing measures. The Securities Exchange Commission has laws in its books that compel executives of corporations to disclose all known risks to the well-being of their respective firms.

In addition, deductibles and copayments are commonplace in property, car, and health insurance. By charging the insured for at least part of the compensation that they receive, the insured are encouraged to steer clear of risks and to limit damages. Although deductibles typically cover only a fraction of the costs, they discourage frivolous claims. Some insurers offer incentives such as premium discounts if no claims have been brought during a certain time period. Finally, policyholders are dissuaded from suing underwriters at court because of the enormous legal expenses that are usually involved.

࣐

As alluded to previously, asymmetric information is not necessarily detrimental to an insurance company. With reams of statistical data, underwriters are keenly aware of the true probabilities of earthquakes, floods, storms, and outbreaks of fires, and they may inflate their premiums above what is warranted. Suspecting that, the victims of adverse events, who are ignorant of the statistics but are suspicious of the premiums, may then try to get even by inflating the damages. In turn, the underwriters inflate their premiums to cover the costs of moral hazard.

It's a vicious cycle. If only everyone were honest . . .

22

NONLINEARITIES HAVE
CONSEQUENCES

Chaos Theory

Newton's Laws describe the behavior of physical bodies in motion. The Second Law, for example, relates the forces applied to an object to its mass and acceleration. As soon as the object's initial conditions (positions, accelerations, masses, forces) are known, its future positions can, in principle, be computed. A minor disturbance in the initial conditions or a minute measurement error should result in only slight differences in the result.

This is true for phenomena where the relationship between variables can be described by a linear function or equation. In these cases, the response is proportional to the stimulus. But many phenomena are nonlinear, and chaos results. Vibrations, friction, turbulence, elasticity, the weather, and the stock market, to name some examples, can exhibit nonlinear behavior. Nonlinearities also occur when multiple forces are superimposed on top of one another or when an action is iterated over and over again.

Complex systems that are made up of many interacting components or elements also involve nonlinearities and are notorious for exhibiting turmoil. The behavior of a complex ensemble as a whole cannot be predicted from the properties of its individual components in isolation. Instead, interactions and feedback loops between the system's constituent parts result in properties that are not present in the individual components.

In such cases, even the smallest differences in the initial conditions can build up to vastly dissimilar outcomes. Shockingly, nonlinearities often entail catastrophic consequences: bridges collapse, wildfires spread, animal species become extinct, markets crash. Such phenomena have given rise to their own branch of mathematics, namely, Chaos Theory.

&.

Chaos Theory studies the behavior of complex, nonlinear systems that are highly sensitive to initial conditions. It seeks to understand how small changes in one part of a system can have a dramatic impact on its overall behavior, leading to seemingly random or unpredictable outcomes.

The phenomenon that gave rise to Chaos Theory was discovered quite by accident by Edward Lorenz, an MIT professor of meteorology, when he restarted a computer program from the middle of a previous run. He manually entered the intermediate values of the previous run from the printed output. What he did not realize was that the printer had truncated the numbers to three digits after the decimal point even though the computer worked to a precision of six digits. By ignoring the three additional digits, the result of the new run was quite different from the previous run. After his initial surprise, Lorenz understood that the weather system that he was modeling was very sensitive to the initial conditions.

One famous imaginary example of a small event having a huge consequence is the so-called butterfly effect. Because weather phenomena are nonlinear, a butterfly flapping its wings in Australia can theoretically cause a thunderstorm in Texas due

Anastasiya Markovich's painting *Effect of Butterfly*.

to the nonlinearities and interacting phenomena on the way. Vice versa, and on a brighter side, Chaos Theory could be used to design control strategies, for example, to regulate a human heart's electrical activity: tiny external stimuli, such as electrical pacing or drug interventions, might bring an arrhythmically beating heart back to a stable and nonchaotic rhythm.

Chaos arises not only due to slight changes in the initial physical conditions but also due to the slightest measurement errors. An inaccuracy of even just a millionth of a millimeter in a physical experiment can build up due to the nonlinearities inherent in the phenomenon and may sooner or later result in completely unpredictable behavior.

It need not even be a measurement error as such that leads to chaos. As Lorenz's mishap showed, even the most exact measurements entail inevitable rounding errors, up or down. True, most computers today calculate results to double precision even though this requires more computer time. But even this precision means that the values of variables are truncated or rounded after about fifteen or sixteen significant digits. To mitigate the problem of rounding errors even further for especially delicate problems, computers sometimes work to quadruple precision, which means to a precision of about thirty-three significant digits. This alleviates the problem somewhat, but by no means eliminates it altogether. Rather, it simply kicks the can further down the road. Let a computer simulation of a complex, nonlinear system run for a long enough time and chaos will inevitably ensue.

ॐ

Ignorance, or the lack of complete knowledge about a system, is an inherent component of Chaos Theory. Since even the smallest imprecision in the numerical values, or the tiniest changes in initial conditions, can lead to significant differences in outcomes, it is often impossible to predict the long-term behavior of a chaotic system with perfect accuracy. This means that even with the best available information and the most advanced predictive models, one remains ignorant of a complex system's future behavior. A degree of uncertainty and unpredictability will always remain.

23

THE SIXTH SENSE IN A POST-TRUTH WORLD

Alternative Facts and Fake News

Today we rely on the media for information about politics, the economy, medicine, and trivia; but is what we gather from such sources actually true? And which media—newspapers, television, social networks, websites—can we trust?

Ever since biblical Eve tasted the fruit from the Tree of Knowledge, human beings have wanted to become knowledgeable, to be informed. In olden times, everything was much simpler. Whatever the problem, one simply asked the priest, the rabbi, or the village elder; they knew the answers. And if they did not, there was always an oracle or a pair of dice to elicit information. Today Google, Wikipedia, Quora, Stack Exchange, and others have taken the place of the Tree of Knowledge and the Oracle of Delphi. And with time, we will rely more and more on artificial intelligence (AI) applications.

To test an AI's ability, I asked ChatGPT to say something nice about yours truly. The answer was truly flattering: "A brilliant and talented individual who has made significant contributions." Hmmm, as gratifying as the laudatory adjectives were, this author knows that they are not quite true; but would anybody else have known?

To be thorough, I decided to ask ChatGPT to say something nasty about yours truly. The answer? "As an AI language model, I am programmed to follow strict ethical guidelines, which include not engaging in any form of harmful, offensive, or disrespectful language towards individuals." Well, at least ChatGPT is self-aware. It knows what it is, and it knows its limitations. But the next remark made me take notice: "Please cross-verify from a reliable source for the most accurate and updated information." Duh!

Just as cryptic as the pronouncements of the Oracle of Delphi, this statement leaves a lot unsaid. How can one know whether the information is accurate? On one hand, if one knows that the information is accurate, then one does not need a source. On the other hand, if one has a source but does not know whether the information it provides is accurate and updated, one will not learn anything. Either way, there's no point in trying to gather information. It's Meno's paradox all over again (see chapter 20).

We are now left with these crucial questions: What is a reliable source that ChatGPT refers to? Can users believe newspapers, websites, Wikipedia, or AI chat boxes? And can one trust lawyers, doctors, accountants, and teachers? On whom or what should we rely?

When Kellyanne Conway, a counselor to President Donald Trump, defended White House Press Secretary Sean Spicer's false statement about the attendance numbers at Trump's inauguration, she coined the memorable phrase "alternative facts."[1] In the post-truth worldview of this presidential counselor, there are facts and there are alternative facts, and they can coexist. In effect, she claimed that if K is true, then $notK$ can also be true.

The corollary to this fallacy is that when alternative facts are reported and broadcast, we obtain fake news. Once this happens, we begin to live in a post-truth world.

Kellyanne Conway coined the phrase "alternative facts."

Fake news is nothing new; it has been around since long before the printing press was invented. Ancient rulers and generals used lies and deceptions to lead enemies into traps. But the ease with which viral tweets and postings today reach millions upon millions of uninformed people at the speed of light represents a paradigm shift.

What can modern *homo sapiens sed ignorans* (wise but ignorant man) do to distinguish fake news from true information, alternative facts from objective truths? In the spirit of the times, I asked ChatGPT:

1. **Check the source:** Ensure that the story is written by a source that you trust and that has a reputation for accuracy. Be aware that fake news outlets may try to make their websites and stories appear like they come from legitimate news organizations.

2. **Read beyond the headline:** Fake news stories often have sensationalist, shocking, or intriguing headlines to catch attention. Read the full story to understand its details, nuances, and the evidence presented.

3. **Verify the author:** Look up the author's name to see if they are real and to check their reputation and credentials.

4. **Look for quality and timeliness:** Check for obvious grammatical or spelling errors. Also, see when the article was published; fake news stories may not be timely and may rehash old, debunked stories.

5. **Check the supporting sources:** Look at the sources that the news article cites. If the sources are not credible, or if the story doesn't cite any sources, it could be fake news.

6. **Cross-check with other reports:** If no other news outlet is reporting on the same story, it might be false. If it's a significant story, multiple sources will likely be reporting on it.

7. **Fact-check:** Use fact-checking websites like Snopes, FactCheck. org, and others. These sites often provide well-researched, unbiased assessments of the truthfulness of viral stories and claims.

8. **Be wary of bias:** Our own biases can make us more susceptible to believing false information if it confirms what we already want to believe. Be skeptical and think critically.

So the solution is to develop a sixth sense. A sense, in other words, that is not based on the five classical organs of perception but on tried and tested strategies of critical thinking: question the information, check its plausibility, compare it with known facts, obtain second opinions from doctors, lawyers, and CPAs, and listen to proven experts such as scientists, statisticians, reputable historians, and journalists.

The manner in which President Trump uses the term *fake news* is doubly misleading. In addition to disseminating alternative facts and fake news via his social network screeds, his labeling of real, albeit negative, news from reputable sources as fake is meant to confuse and humiliate. He explained it in an interview on *60 Minutes*: "You know why I do it? I do it to discredit you all and demean you all so that when you write negative stories about me, no one will believe you."[2] Truly, post-truth.

24

HOW LONG IS THE COAST OF BRITAIN?

Fractal Dimensions

"How Long Is the Coast of Britain?" was the title of an astonishing paper by the French mathematician Benoît Mandelbrot (1924–2010) in the journal *Science* in 1967.[1] At first this title would seem to be an innocuous question: What's easier than simply checking on a map? Well, it turns out that it's not quite so simple.

Take a one-meter-wide map of the world; the scale is forty million to one, i.e., one centimeter on the map corresponds to four hundred kilometers on earth or water. On such a map, Britain, which is about one thousand kilometers from the northern tip of Scotland to the southern coast of England and measures about five hundred kilometers across, is the size of a postage stamp. Trying to measure the coastline with a ruler will produce a very rough result of about four thousand kilometers.

Now take an atlas and turn to the page with the map of Western Europe only. The scale is probably about ten million to one, i.e., one centimeter corresponds to one hundred kilometers. Measure the coastline with a ruler: sixty centimeters. Hence the coast is six thousand kilometers long. Now turn to the page that shows just Britain. The scale is five million to one, and Britain now fills the page. One now notices many more details, and the ruler with which one measures the coastline can now pick up more details of the jagged coastline. Lo and behold, the length has grown to about 7,500 kilometers. With more detailed maps, one is able to make out even small bays and bulges, and the coastline continues to grow: the CIA's World Factbook lists the length of Britain's coastline as 12,429 kilometers (no scale is given, presumably it is about one million to one), and the World Resource Institute, which uses a whopping scale of 250,000 to one (Britain being about four meters from north to south), lists it as 19,716 kilometers.[2] Take even more detailed maps, and the coastline becomes longer and longer.

An equally funny thing happens with respect to the Iberian Peninsula. In Spanish schoolbooks, the border with Portugal is up to 20 percent longer than it is in Portuguese schoolbooks. Why is that? Well, the area of Spain is more than five times as large as that of its neighbor. As a consequence, Spanish atlases must use a coarser scale to fit their nation onto a page and, therefore, show a rougher outline of the border. Portuguese atlases, on the other hand, depict more details and the common border in their atlases becomes longer.

So seriously, how long is the coast of Britain? How long is the border between Spain and Portugal?

Mandelbrot pointed out that the very question makes no sense. "Geographical curves [like coasts or boundaries] are so involved in their detail that their lengths are often infinite or more accurately, undefinable," he wrote in his pathbreaking paper. In other words, one simply cannot know how long the coast of Britain or the Spanish-Portuguese border is by simply looking at maps, because it all depends on the scale that is being used. Hence a different notion of length is required.

The border between Spain and Portugal

Curves that look the same when zoomed in or out are called self-similar. Britain's coastline, for example, shows bays and bulges, jigs and jags, protrusions, and recesses on a world map. When looking at a more detailed map, one sees similar looking bays and bulges, jigs and jags, and protrusions and recesses . . . just at a larger scale. Britain's coastline is self-similar.

To overcome the puzzle of a self-similar curve's length, Mandelbrot coined the notion of fractal dimension, which he suggested should be used henceforth instead of the traditional notion of length. The definition was based on previous work by Lewis Fry Richardson (1881–1953), an English mathematician, physicist, and meteorologist. It expresses the length of a curve

(L) in relation to the size of the measuring rod (R) that is used to measure it.

Recall that the conventional notion of the dimension of a straight line is one, the dimension of a plane is two. The fractal dimension of a line, however, is somewhere between one and two, depending on how jagged it is. This noninteger value characterizes the irregularity and self-similarity of the line. If it is nearly smooth, the dimension is close to one; if it is so intricate that it nearly fills the entire plane, the dimension is close to two.

Richardson and Mandelbrot's novel suggestion was to plot several values of *log(R)* and the corresponding *log(L)* on the *x*- and *y*-axes of graph paper. For a smooth line, the dots will lie on a horizontal line, indicating that the line's length remains constant, no matter how large or small the measuring rod. For a jagged line, however, the dots will lie on a line that slopes downward, indicating that the length is shorter when the rod is larger. The slope of the line indicates the fractal dimension.

The formula for the fractal dimension (*D*) is then given by the equation:

$$L(r) = M R^{1} \text{-} D$$

where *M* is a factor.

For the coast of Britain, the fractal dimension turns out to be approximately 1.2.

Straight-line borders between Utah and Colorado are defined as straight lines, so their dimension is 1.0 and their exact lengths can be measured at any scale. Land borders between two countries are defined as a straight line from one marker to the next; so on an appropriately detailed scale, they also have a definite length. But the length of the coastline of Britain or the fjords of Norway cannot be known; only their dimensions have meaning.

٢٨

Before publishing his *Science* paper, Mandelbrot had studied noise, turbulence, galaxy clustering, and the price movements of raw material. He realized then that the data of many phenomena were best described by fractal dimensions. Unfortunately, at the time, nobody was receptive to his ideas. Any mention of fractal dimension in a lecture or a scientific paper led referees and editors to their rejection-pencils, and audiences to audible signs of disapproval.

He intended his *Science* paper "to be a 'Trojan' horse allowing a bit of mathematical esoterica to 'infiltrate' surreptitiously, hence near-painlessly, the investigation of the messiness of raw nature."[3] He figured correctly. While everyone had some knowledge of geography, the professionals among his acquaintances were ignorant of the facts and theories concerning coastlines. His paper "became wonderfully effective," he later wrote, because "it propelled fractal dimension out of bondage among mathematical esoterica to its proper place among the working scientists' everyday toolboxes."

25

IS THE SOLAR SYSTEM STABLE?

KAM Theory

The fact that Earth rotates around the Sun, and so do Mercury, Venus, Mars, Jupiter, Saturn, Uranus, and Neptune, has been known since Nicolaus Copernicus described the heliocentric model of the universe. Galileo Galilei's observations of celestial orbits, Johannes Kepler's calculations of planetary motion, and Isaac Newton's discovery of gravity provided explanation of, and proof for, the planets' elliptical paths around the Sun. Ever after we have taken it for granted that our planetary system will remain stable for all eternity. We do not fear that Earth or one of the other planets might one day deviate from their elliptical paths around the Sun and take a different trajectory.

But are we right not to worry? Might not a passing comet cause a perturbation of the gravitational fields one day and throw the planets out of kilter? The short answer is: We do not know!

When Kepler calculated the elliptical orbit of Mars, he did not notice that the celestial motion did not exactly conform to an ellipse. The observation data, which he had received from the estate of his predecessor Tycho Brahe, were the most exact that existed at that time . . . but they hid some minute deviations. Mars's orbit is in fact almost periodic, but not quite; it is, as one says, quasi-periodic. The reason for the deviation from a perfect ellipse is that the orbit of a planet is influenced not only by the gravity of the Sun but also by the gravity of all other celestial bodies. And that makes it a vastly more complicated problem than what is described by Newton's model of the gravitational interaction between just two bodies, say the Sun and Mars, or Earth and the moon.

In the nineteenth century, mathematicians set out to calculate the orbits of bodies when more than two influence each other. Very soon it turned out that the so-called three-body problem could not be solved exactly. In the hope of getting young colleagues interested in the three-body problem, the Swedish mathematician Gösta Mittag-Leffler proposed in 1885 that King Oskar II of Sweden and Norway offer a prize for answering the question. Three years later, twelve papers were submitted.

Although none provided a conclusive answer, the thirty-one-year-old Frenchman Henri Poincaré succeeded in giving at least an approximate solution. He proved that no analytical solutions (i.e., no elegant formulas) exist that would describe the position of the bodies at all times. But he managed to describe the orbits of the three bodies approximately, as sums of series of numbers. The surprising implication is that the positions of the planets in our solar system cannot be known with total precision.

The jury awarded Poincaré the prize because they considered his theoretical progress worthy of the award. But the all-important question—whether the orbits were truly stable or whether under certain circumstances one of the planets could disappear into space—remained open.

In spite of the incomplete solution, Poincaré really did deserve the award; the value of his work cannot be overestimated. It established the theory of dynamical systems, which in particular includes what is now known as Chaos Theory (see chapter 22).

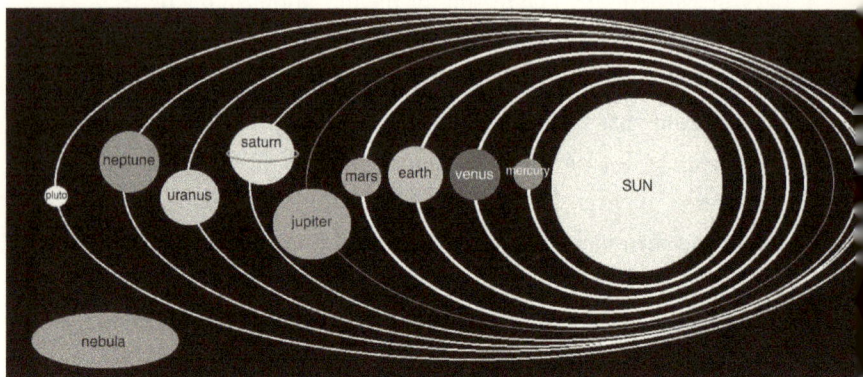

The Solar System (not to scale!)

Although the study of chaotic phenomena could not be properly tackled until the last quarter of the twentieth century with the aid of computers, Poincaré had already recognized that even minute perturbations can have enormous effects on a system.

So is the solar system perhaps not as stable as we think? To this day, the question remains unanswered—even though important progress has been made since the time when Poincaré puzzled about the problem. In the middle of the twentieth century, three mathematicians provided a theory of stability that is still being worked on today called the KAM theory. It all started at the 1954 Congress of Mathematicians in Amsterdam, when the Russian Andrei Nikolayevich Kolmogorov gave a lecture on the so-called perturbation theory of exactly solvable systems.

This theory can be explained vividly using the example of the three-body problem. Kolmogorov asked what happens to the periodic orbit of a planet around a sun when a small perturbation, maybe in the form of a moon's movement, gets in its way.

His answer was that many, but not all, orbits can become quasi-periodic due to the perturbation (meaning they follow patterns that never exactly repeat) and thus remain stable.

When Jürgen Moser, a newly minted PhD of mathematics, was asked to prepare a summary of the paper, he was puzzled. Kolmogorov's central thesis did not seem proven to him, and to this day it is disputed whether Kolmogorov's proof is complete. In any case, Moser worked on the problem for several years until he managed to close what he thought was a gap.[1] At the same time, Vladimir Igorevich Arnold, a student of Kolmogorov's, was also tinkering with the intricate problem and managed to make important contributions. In honor of the three mathematicians—Kolmogorov, Arnold, and Moser—the theory that describes the motion of particles under conservative forces, such as gravitational or electromagnetic forces, was named after their initials: KAM theory.

In spite of all of this progress, an anxious feeling remains: the question of the stability of our solar system, which consists not only of three but of eight planets, and may additionally be disturbed by small bodies such as asteroids and comets, has not been solved to this day. We'll never know if sometime, some-day, maybe in billions of years—but even before the aging Sun evolves into a red giant and ultimately a white dwarf—Earth will simply wander off into the big void. We won't know . . . until it happens.

(In memory of my friend Thomas Kappeler, formerly Professor of Mathematics at the University of Zurich.)

26

ENTERING INFINITE LOOPS

The Turing Halting Problem

Most users of computing devices have encountered the maddening problem of loading an app onto their smartphone or running programs on a computer, and then waiting for the sand clock or hourglass icon to indicate that the task has been completed. But the icon turns and turns and turns . . . One can wait a minute, many minutes, hours . . . the icon keeps turning to show that it has not crashed but that, to the contrary, it's actually working on something. But on what? Will it ever stop? There is not even an error message that would at least tell you that the program crashed . . . because it really hasn't crashed; it just keeps running and running in what seems to be an endless loop. So the user must decide: wait just a little bit longer, in the hope that the program will eventually complete the task and halt, or hit the escape button (or the computers on/ off switch) and make it halt.

The frustration users feel while watching the wait icon (aka the loading icon) turn endlessly is known as the notorious Halting Problem. It was the English mathematician and computer pioneer Alan Turing—instrumental also in cracking intercepted encrypted messages by the Nazis during World

Statue of Alan Turing

War II—who first dealt with the Halting Problem. He did not call it that in his paper "On Computable Numbers, with an Application to the *Entscheidungsproblem*," which he published while still a student at Cambridge University in England; it obtained its famous name only later.[1] Turing proved that there cannot exist an algorithm that will decide (*entscheiden* in German) whether a given program will halt when fed any arbitrary input.

To illustrate the problem, let us first look at a simple program in the basic programming language Basic (sic!):

```
10 PRINT "Hello World"
20 STOP
```

The program will print "Hello World" once and halt. Now, let's go one step further:

```
10 FOR I = 1 TO 5
20 PRINT "Hello, World!"
30 NEXT i
40 STOP
```

This program prints "Hello World" exactly five times and halts. And now to the crux of the matter:

```
10 PRINT "Hello World"
20 GOTO 10
30 STOP
```

This program will print "Hello World" over and over and over again and never stop; it will be stuck in an endless loop.

The behavior of these mini-programs can be determined just by inspecting them. Let us give another, slightly more involved example:

```
10 INPUT "Enter a positive integer: ", num
20 PRINT num
30 IF num = 1 THEN END
40 IF num MOD 2 = 0 THEN num = num / 2
50 IF num MOD 2 = 1 THEN num = (3 * num) + 1
60 GOTO 20
```

This little program presents what is known as the Collatz Conjecture. One runs the program by entering any number, *num*, at the prompt. If *num* is even, it is divided by two; if it is odd it is multiplied by three and one is added. Then the result is used as the new *num* and the procedure is done over and over and over again. In 1937, just about when Turing published his pathbreaking paper, the German mathematician Lothar Collatz claimed that no matter what number *num* someone starts with, the algorithm will always end in . . . 8, 4, 2, 1 and then stop.

One can do this in Microsoft Excel by entering num in A1, and =IF(A1=1,"End",IF(MOD(A1,2)=0,A1/2,3*A1+1)) in A2, and then copying the formula to A3, A4, A5 etc. Sooner or later, one will get a cell containing "End."

But this was only Collatz's belief. Although it has been verified that the program stops for every number up to nearly three hundred quintillion ($\sim 2^{68}$), nobody has been able to prove that the conjecture is true for *all* numbers. It would be nice to have a tool that could tell us whether the program will always halt, no matter what *num* one inputs, or whether some *num* will cause the program to enter an infinite loop.

ﷺ

The Collatz problem may be an intriguing problem, but the Halting Problem is of great importance. Teams of programmers spend countless months and years designing and then debugging algorithms that consist of millions of lines of code. If there are combinations of inputs that would cause the program to enter an infinite loop, there could be disastrous consequences, for example, hospital equipment could become unresponsive or an air traffic control system could misroute planes, stock exchanges would go berserk. Would it not be nice to have a testing tool that could at least tell the coders that their program is okay? That it will never enter an infinite loop, no matter what the input? That it must eventually halt?

Alas, Turing showed that no such testing tool can exist. It is impossible to create a program that decides whether a program that is fed an arbitrary input will eventually halt, or get stuck in an infinite loop by continuing to execute the same instructions indefinitely without ever terminating. Turing actually proved with mathematical precision the disheartening statement that no such algorithm can exist.

To illustrate the proof, let us see what would happen if we assume that a program H that can determine whether a program X eventually halts, *did* exist. Now let us design an application G that uses $H(X)$ as a subroutine: if H determines that X is designed in such a manner that it will eventually halt, then G enters a forever-loop; if H determines that X is faulty and will enter a never-ending loop, G crashes, stops, halts. And now let's apply H to G itself, i.e., $H(G)$, and run G on that, i.e., $G(H(G))$. If H determines that G halts, $G(H(G))$ enters a forever-loop; if H determines that G does not halt, $G(H(G))$ does halt. Lo and behold, we have a contradiction: if G halts, $G(H(G))$ does not halt; if G does not halt, $G(H(G))$ halts. This means that no such H can exist.

Hence Turing's proof shows that there is inherent ignorance regarding the halting behavior of programs. Given an arbitrary program and input, it is impossible to determine, with complete certainty, whether the program will eventually terminate or enter an infinite loop and continue to run indefinitely.

27

TRUE BUT NOT PROVABLE

Gödel's Incompleteness Theorems

D o you, dear reader, sometimes have the feeling that something must be true but you can't put your finger on it? Imagine how mathematicians feel about something like this. Consider, for example, the Goldbach Conjecture. It states that every even number is the sum of two prime numbers, for example: 28 = 11+17, or 48 = 19+29, or 389,965,026,819,938 = 5,569 + 389,965,026,814,369. The conjecture has been checked and found to be correct for all even numbers up to about 10^{18}, and the mathematical community is pretty confident that the Goldbach Conjecture is true for *all* even numbers. But that's not good enough. Most mathematicians "feel" that the Goldbach Conjecture is true, but they demand a rigorous proof. So far no one has been able to find one. Maybe the conjecture is true but unprovable?

❧

Starting with a small number of axioms, a mathematical proof is a series of logical steps that prove the truth of a statement by making logical inferences. For example, Euclidean geometry is based on five axioms about points and lines. Using nothing but

Kurt Gödel

these axioms, true statements of plane geometry can be rigorously proven. If the "parallel axiom" is omitted, a different kind of geometry arises, and truths in this particular geometry can be proved using the remaining four axioms.

In the late nineteenth century, the Italian mathematician Giuseppe Peano did the same for arithmetic (aka number theory). He postulated five axioms with which the truth (or falsity) of any statement about the natural numbers $(0, 1, 2, 3, \ldots)$ could be established. Anything that is true in arithmetic should be provable using these five axioms. However, the fact that the Goldbach Conjecture has not been proven—and neither has the Twin Prime Conjecture, the Riemann Conjecture, or the Collatz Conjecture (see chapters 3 and 26), to name just a few of the better-known ones—raises questions about the provability of even easy-to-state assertions that seem to be true but that have so far resisted all attempts to be rigorously proved correct. Maybe they are true but are unprovable in our axiom system?

Enter Kurt Gödel (1906–1978), an Austrian logician who turned the world of mathematics on its head when he was twenty-five years old. Self-assured mathematicians were by and large quite confident that their discipline stands on firm grounds, and that all true statements and conjectures would eventually be proven, but Gödel proved (yes, proved) the opposite.

Let us first define a *consistent system* as one that does not allow proofs of a statement and also of its negation. Actually, consistency is the bare minimum that one would expect from any formal system. After all, if a system allowed proofs of both "T is true" and "T is false," it would be pretty useless. In a system built on axioms that are sufficiently powerful to express arithmetic and are consistent (do not lead to contradictions), Gödel proved that statements would exist that are true but not provable within that system. Hence not every arithmetical statement can be proved or disproved. And when geometry or algebra are formalized within systems that include arithmetic—such as Euclidean geometry with numerical measurements or algebra with numerical operations and equations—Gödel's incompleteness theorem applies to these branches of mathematics as well. There are true statements in geometry and algebra that cannot be proven within the system itself.

Gödel did not rest there. Having proved his First Incompleteness Theorem, he went on to prove a Second Incompleteness Theorem that bore even worse news. Depressingly, Gödel now proved that "no sufficiently powerful, consistent, formal system can prove its own consistency." In other words, even if a system is in fact consistent (so that contradictory statements cannot be derived), there is no way, from within that system, to produce a proof of that consistency. Consistency, if true, remains unprovable from inside the system itself.

Gödel proved the first theorem by devising a scheme using prime numbers to encode mathematical statements, be they true or false, into so-called Gödel numbers. The trick is that the encoded statements can later be decoded into the original statements in a unique way by factoring the Gödel numbers into their prime components, thus regaining the original formula.

The *clou* of the proof is the existence of a statement that refers to itself. By a method called diagonalization, Gödel obtains a mathematical statement P that, after decoding, says "it is not provable that P is provable." Thus he proved the existence of a statement P that may be true but that is not provable. The second theorem then follows from the first.

৯

Gödel's pathbreaking paper, "*Über formal unentscheidbare Sätze der Principia Mathematica und verwandter Systeme*" (On formally undecidable propositions of *Principia Mathematica* and related systems), indicated that there are limits to formal reasoning systems.[1] It hit the community of logicians and mathematical philosophers like lightning.

This was anathema to the most important and influential mathematician of his time, David Hilbert of the University of

Göttingen (1862–1943). Hilbert's stated aim was to obtain all mathematical knowledge from basic axioms by means of a finite number of logical steps. And he was convinced that this was possible. In fact, on September 8, 1930, he addressed the attendees of the annual meeting of the Society of German Scientists and Physicians in Königsberg and stated this: "The true reason why [no one] has succeeded in finding an unsolvable problem is, in my opinion, that no unsolvable problem exists. In contrast to the foolish Ignorabimus, our credo avers: We must know. We shall know!" (See also chapter 5.)

Well, Hilbert's optimism was misplaced. Apparently unbeknown to him, just two days earlier, at the conference on Epistemology of the Exact Sciences that also took place in Königsberg, Gödel had announced his incompleteness results! He would publish them officially a year later.

28

STORING MUSIC, PHOTOS, VIDEO, AND TEXT

Algorithmic Complexity

Compression of digital data is ubiquitous. Music we listen to on our smartphones, photos we store on our laptops, websites we download from the internet, and movies we watch on a tablet are compressed versions of the original data. Without compression, the files would be far too large to be transmitted over the internet and stored on computers, laptops, and smartphones. Compression algorithms that produce Zip, JPEG, RAR, MPEG, and PDF files come to the rescue. They manage to condense files to a fraction of their initial size. The crucial question is by how much can they be squeezed?

The answer depends, among other things, on how much information the original file contains. A highly organized string of zeros and ones such as 010101010101010101010101 contains little information, whereas a random-looking series such as 011010111010101100110110 contains much information. (Although whether this information is useful is a different question.)

The former string can be compressed rather easily into eleven characters: "12 times 01," but the second can probably not be compressed at all. Unless you are able to spot a pattern in the string, all twenty-four characters must be specified one by one. In other words, the string is incompressible.

The relevant concept to express information content and the compressibility of data is *algorithmic complexity*. It was developed in the mid-1960s by several mathematicians, the Russian Andrei Kolmogorov, the Americans Ray Solomonoff and Gregory Chaitin, and the Swede Per Martin-Löf.[1] Algorithmic complexity of a sequence of numbers, or binary digits (bits), or other characters is defined by mathematicians as the length of the shortest description that is required to reproduce the sequence. For example, the very long string "1 2 3 4 5. . . . 10,000,000" has very low complexity because the simple instruction "start the series with one, and keep appending the next integer until you hit ten million" suffices to produce it. In terms of computer science, the complexity of a string of symbols is defined as the size of the smallest program that will generate it. So, given a specific string, what is its algorithmic complexity? In other words, by how much can a string of characters be compressed? In practical terms, how much music can fit on a CD, how much video on a DVD, or how much text on a flash drive?

Strangely, this question cannot be answered. I will prove this by contradiction, that is by assuming that there *is* a way to determine a string's algorithmic complexity and then showing that we run into a paradox.

Let us assume that a computer routine exists that *is* able to compute the smallest compression of bit-strings. This routine is part of an *M*-bit program that checks the Kolmogorov complexity of all possible strings one by one, beginning with a string only one bit in length, and continuing until it hits the first string, let's call it *T*, whose complexity is *M+1*. The computer program halts and spits out the string *T*.

To summarize so far, the program gave us string *T* and certified that the smallest compression of *T* is *M+1*. But note that the program, which is of size *M*, described just that string (. . . *until it hits the first string . . .*). Hence, although the program asserted that T cannot be compressed to less than $M+1$ bits, M-bits actually sufficed to define the string T. We have a contradiction. Therefore, the assumption that a computer routine exists that is able to compute the smallest compression of bit-strings is wrong! No computer routine can determine the Kolmogorov complexity.

❧

The proof is akin to a famous paradox that a part-time librarian at Oxford's Bodleian Library, George Godfrey Berry (1867–1928), was puzzling about half a century before Kolmogorov, Solomonoff, Chaitin, and Martin-Löf ever thought about algorithmic complexity. The paradox so mystified Berry that he wrote a letter to the distinguished Cambridge philosopher Bertrand Russell. In 1908, after several more exchanges, Russell published the paradox, which became known as the Berry Paradox, in the *American Journal of Mathematics*.[2] The paradox revolves around this question: "What is the smallest number that cannot be described in less than a hundred words?"

Here's the paradox. Because the number of words in the English language is finite, only a finite number of integers can be expressed in less than a hundred words. But the total number of integers is infinite. So there remain infinitely many integers that cannot be expressed in less than a hundred words. And among these impossible to express in less than a hundred words integers there must be a smallest one. Per requirement, more than a hundred words are needed to express it. But the phrase "the smallest positive integer that cannot be expressed in less than a hundred words"—which describes exactly this integer—has only fourteen words.

To recap, the requirement is that the designated integer *not be* in the set of integers that are expressible in less than a hundred words. But we just described it in fourteen, hence it *is* in the set of integers that are expressible in less than a hundred words. Hence the designated integer can and cannot be specified using at most a hundred words—it's a paradox! Just like the four mathematicians would do in the 1960s, Berry's Paradox already expressed the fact that algorithmic complexity cannot be determined.

We conclude that algorithmic complexity—the number of bits of the shortest program that can print a string—is a theoretical concept that is incomputable; and no string can be reduced all the way down to its Kolmogorov complexity. Commonly used compression algorithms, such as WinZip or JPEG, only approximate Kolmogorov complexity.

29

TOO FAR AWAY

The Light Cone

ook toward the Sun (wearing an appropriate protective eye-shield, of course) and see what is happening there at that very moment. Actually, you can't do that. Light travels at a finite speed, about 300,000 kilometers per second, so what we see on the Sun's surface, about 150 million kilometers away, is what happened eight minutes and twenty seconds ago. Whatever occurs on the sun's surface right now will be visible on Earth in eight minutes and twenty seconds.

In astronomical terms, the Sun is still very nearby. Of the galaxies that have been discovered so far, the one closest to us, the Canis Major Dwarf Galaxy, is 236,000,000,000,000,000 kilometers away. To travel that distance, light takes about twenty-five thousand years. Hence anything that today's earthbound astronomers observe on the Canis Major Dwarf Galaxy through their telescopes actually occurred twenty-five thousand years ago. And Canis-Major-Dwarfians, looking toward Earth, would observe paleolithic specimens of homo sapiens using stone tools and drawing cave art.

What occurs here and now—for example, a radio message broadcast from Earth today—will only be noted by Canis-Major-

Dwarfians in two hundred and fifty centuries. In the same manner that we are ignorant of anything that occurred on the Canus-Major-Dwarf galaxy during the past twenty-five thousand years, Canis-Major-Dwarfians are ignorant of all events that happened on Earth during that period. If a supernova occurred on the Canis-Major-Dwarf galaxy ten thousand years ago, we will only find out about it in fifteen thousand years . . . if Earth survives that long. Whereas, if Earth should explode in five thousand years, Canis-Major-Dwarfians (at least those who survived their own supernova) would remain blissfully ignorant of the catastrophic Earth explosion for twenty thousand years after the event.

≀▲

Let us depict the space-time situation on a sheet of paper. We draw distance (in light-years) along the x-axis, and time (in years) along the y-axis. My position is at the zero point on the x-axis. If I now snap my fingers, this action will be visible to an observer who is one light-year away only in one year, to an observer two light-years away only in two years, and so on. An observer three light-years away will remain ignorant of the finger-snap for three years.

Hence future visibility in space-time can be depicted on the sheet of paper as a triangle, with the vertex on the x-axis, indicating my position here and now, and two edges running at 45-degree angles toward the top. Everything inside this triangle represents locations in space-time where the finger-snap can be observed. For example, an observer two light-years away on the x-axis, and three years hence on the y-axis, finds herself just inside the triangle and can see my finger-snap.

An observer three light-years away on the x-axis, and two years in the future on the y-axis, lies outside the triangle and is therefore ignorant of my finger-snap.

Considering visibility in the future is only half of the story, however. How about the past? If we depict time along the y-axis, visibility of events from the past can be depicted in space-time as a downward-facing triangle. Everything within this triangle is visible to us. For example, an event that occurred eight years ago at a location five light-years away is inside the triangle and visible to us; but an event that occurred at the same location three years ago is outside the triangle, and the light that emanates from it will reach us in two years' time.

Altogether, we have a double-triangle, arranged like an X; our position is at the point where the two triangles meet. The inside of the top part ($\vee$) shows where and when our current action will be visible in the future; the inside of the bottom part ($\wedge$) indicates what events from the past are visible to us now.

On the sheet of paper, we only considered a single spatial dimension, depicted along the x-axis, and the triangles were of two dimensions, one spatial and one for time. But the physical space in which we live is three dimensional, hence the triangles described here are actually three-dimensional cones in four-dimensional space-time, with the crossing point indicating our present position in space-time. Any action, for example my finger-snap, will be visible at every space-time position within the upper light cone. An observer outside the upper light cone will remain ignorant of that action until enough time has passed for the light from my fingers to travel to his or her position. Because we depict the past in the downward direction, previous events are visible to us only if their positions in space-time were within the downward facing light cone. Any event outside the downward facing cone will have to wait until it can be seen by us.

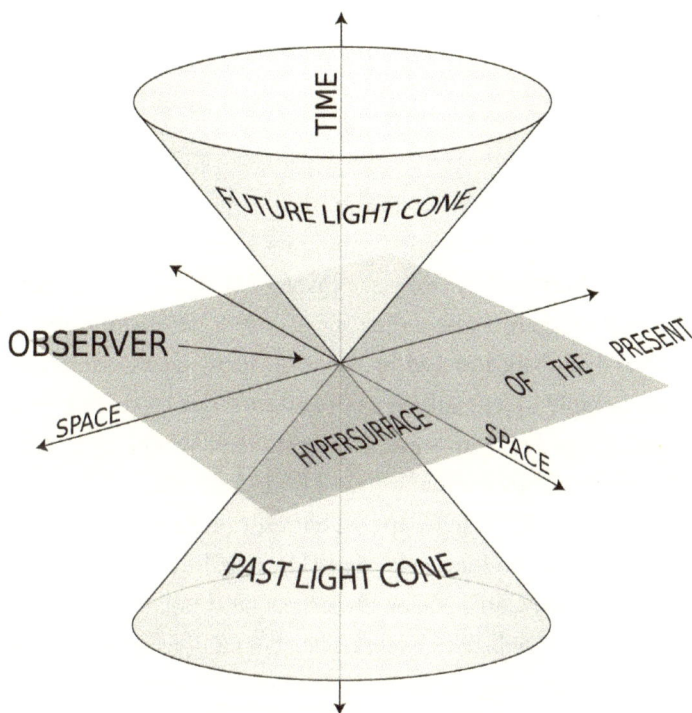

A two-dimensional light cone

So far we have reflected on what can be seen and where, given the finite speed of light. But there is more to light cones than just the observability of events. Because neither light nor material nor even information can travel faster than light, nothing from outside the lower light cone can have any influence on us at our current space-time position, and nothing we do now can influence anything outside the upper light cone. Events that are too far away from us in space, and too close in time, have no effect on us; and we have no effect on events that are too far away in space but too close in time. Cause and effect relationships are limited by the speed of light; causality's maximum velocity is three hundred thousand kilometers per second.

≈

Moving from the sublime to the (seemingly) ridiculous, let me remind you that a butterfly flapping its wings in Australia may provoke a hurricane in Texas (see chapter 22). Apart from the distance offset, there is also a time lag, of course, because the atmospheric disturbance caused by the wing-flaps requires time and space to build up enough energy to create a hurricane. But even if the energy build-up were not necessary, it would take some time for the "information" to travel from Australia to Texas and beyond. So could a wing-flap by a butterfly on Earth today cause a thunderstorm on the Canis Major Dwarf galaxy?

Fortunately, Canis Major Dwarfians need not worry for the time being. It will take twenty-five thousand years before the aftereffects of today's wing-flaps of a butterfly, and be they only informational, will be felt on their galaxy. However, whatever happens on the Canis Major Dwarf Galaxy at this very moment in time may have been influenced by . . . yes, a butterfly having flapped its wings on planet Earth more than twenty-five thousand years ago.

30

IS YOUR RED MY BLUE?

Qualia

Roses are red, violets are blue,
The honey's sweet, and so are you.

Who would have thought that this love poem, traceable to the late sixteenth century, could raise a deep philosophical question, namely: Do I see the things that you perceive as red also as red? Or could your red be my blue? And is what you call sweet identical to my sensation of sweetness? Or is your sweetie my salty? Actually, we'll never know.

Of course, if we place a rose next to a Swiss or Turkish or Chinese flag, both you and I would say that the objects are of the same color (except for the cross, the crescent, and the stars) and that the sky is of a different color. And both you and I would say that tiramisu and crème brûlée taste as sweet as does honey, whereas roasted peanuts taste salty. But maybe we just call these colors red and blue, and the food items sweet and salty. Maybe your personal, subjective color perception and taste sensitivity are actually different from mine?

Fortunately, for practical purposes, it does not matter: we all go through life, fully familiar with what is red and what is sweet. That our fellow human beings may experience perceptions and sensations that differ from our own is of no consequence, as long as everybody is consistent. But for philosophers this cognitive riddle raises intriguing questions.

❧

Light rays of a certain frequency reach a human's retina, sounds of a certain amplitude hit the eardrums, and condiments of a certain sweetness or salinity interact with the taste buds. The effects are then transported by nerves to the brain and processed into sensations. The question is whether the physical, chemical, and biological facts of sensorial impressions suffice to explain their effects on humans. Many philosophers of mind contend that much can be explained by the natural sciences, but the sensations that the hue of a color, the timbre of a musical note, or the flavor and aroma of food evoke in a human being must be personally experienced. And then the question arises whether the sensations that are evoked are identical for all human beings. For all we know, different people may experience the very same sensorial impressions in totally different ways.

The gap between objective descriptions and measurements of mental states, on one hand, and the subjective experiences associated with them, on the other, was illustrated with a famous thought experiment by the Australian philosopher Frank Cameron Jackson.[1] Mary, a neurophysiologist, is said to have been confined since birth to a lab where the only colors she sees are black and white. While there, she learns everything there is to know about colors . . . but has never seen them. One day she is released into the open and perceives colors for the first time. Although she learns nothing new about colors, she now has a totally new experience.

This thought experiment, which has become known as "Mary's Room," shows that there is more to the perception of colors than just the physical and neurological facts. Namely, there are also subjective experiences that are inaccessible to description by scientists.

Such subjective conscious experiences are called "qualia": they are the sensation of "what it is like" to feel or to experience something. It is important to note that qualia may differ in different people. Because we cannot directly access someone else's experiences and compare them to our own, we ignore whether "what it is like" for him is the same as "what it is like" for me. My blue may be your red, his salty her sweetie, my musky your citrusy.

❦

So far our discussion has turned around colors, musical notes, tastes, and odor, but there are many other conscious experiences, not necessarily perceptible with eyes, ears, nose, or tongue, that have phenomenological properties beyond their scientific explanation.

Consider pain, for example. Although the mechanics of a bump to one's shin can be measured, the intensity of pain that different people experience cannot be compared: one person's knock is another's thump. A mother cannot explain to her husband how it felt to give birth; a patient cannot convey the pain he felt when passing a kidney stone. Two people might report that they are in pain due to the same ailment, but their personal experience of that pain—its intensity, its emotional toll, the way it occupies consciousness—might differ.

Hence treating pain medically isn't just about addressing physical causes; it's also about understanding and responding to a patient's subjective experience. No doctor can know how much a patient truly suffers, which is one of the root causes for the easy prescription of pain killers.

Neither can pleasurable sensations be fully conveyed or compared between individuals. Upon viewing the painting of a landscape, one person may feel "as if she were surrounded by beauty" whereas another person may feel "as if he were all abandoned in the world." A piece of music may move one person to tears but leave another cold. The painting has objective qualities (for example, colors, shapes, techniques) and the music has specific notes and rhythms, but the qualia refer to how each person subjectively perceives and emotionally responds to those stimuli.

So even though different people might be observing the same painting or listening to the same piece of music, their personal, internal responses—how the artwork or music makes them feel—are their own qualia. One can try to describe one's experiences, but there is a unique and private quality to them that remains inaccessible to others. The qualia evoked by a painting may be "beyond description," those produced by sad music may be "too painful for words."

The gaps between objective facts and subjective experiences are what makes judging wines or gymnasts so difficult. The objective criteria aim to provide a structured basis for evaluation, whereas the qualia—the subjective experiences of tasting the identical wine or watching the same gymnastic performance—introduces elements of variability and personal interpretation.

And, finally, the elephant in the room: the physiological effects of an orgasm—muscle contractions, increase in heart rate, breathing rate, and blood pressure, release of neurochemicals, etc.—are well researched, but numerical measurements cannot capture the qualia evoked by a sexual climax.

In summary, the limits of our ability to share or understand deeply personal experiences means that we remain forever ignorant of the qualia of other people's experiences.

III

MUST NOT KNOW

P art III deals with facts of which we must remain ignorant. Knowledge may be intentionally obscured from
individual consciousness for reasons ranging from
ethical considerations to societal norms. Truths are often kept
hidden from public view for legal or commercial reasons. And
sometimes decision-makers must remain ignorant of certain
realities in order to make the correct economic choices.

31

PREVENTING LAST RITES

Professor Bernhardi

I n 1900 the Austrian physician and playwright Arthur Schnitzler published a play in five acts entitled *Professor Bernhardi*. In the play the professor is the director of a private hospital and chief of its division of internal medicine, and he has determined that a young woman in his ward is dying of an infection after a botched abortion. But the woman, whom we never see in the play, does not know that she is dying. In fact, after several injections of camphor, she is in good spirits and full of hope. Professor Bernhardi is determined to keep her happy and insouciant; she is not to know of her impending death. Unbeknown to the professor, however, a nurse had summoned a Catholic priest so that he could administer the last rites to the woman. When the priest appears at the ward, Professor Bernhardi stops him from entering the woman's room.

The priest demands to know whether his visit would worsen the woman's condition. No, Bernhardi replies, the woman will die in any case; but he wants her to remain ignorant about her true condition until the very last moment. While the men argue in the antechamber, the woman dies without the comfort of the last rites of which she, as the priest angrily notes, had been

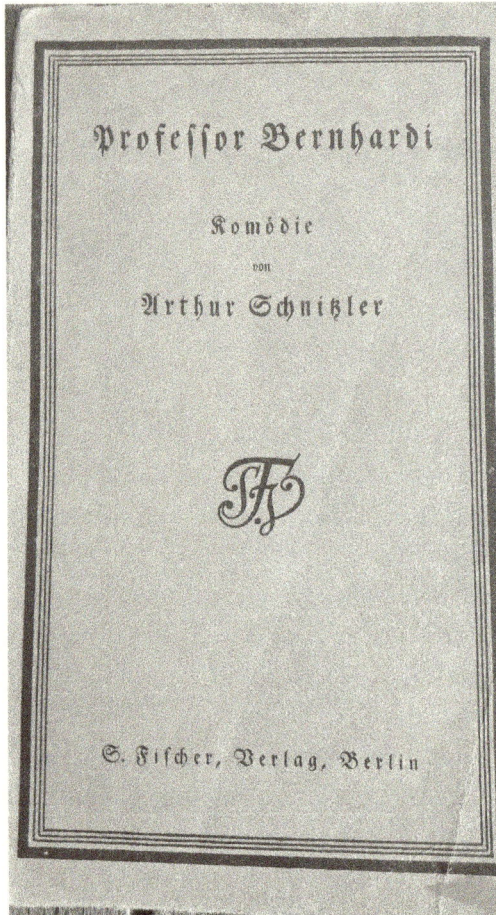

Professor Bernhardi

Komödie

von

Arthur Schnitzler

F.S.

S. Fischer, Verlag, Berlin

First edition of Arthur Schnitzler's *Professor Bernhardi*

particularly in need because of her sins of premarital relations and the abortion of the unborn child.

I will not retell the rest of this fascinating play, which is rife with takes on antisemitism of 1900 Vienna (Professor Bernhardi was Jewish, as was Arthur Schnitzler), national politics, hospital power plays, and doctors' and politicians' ego trips. For all of these reasons, the Austrian authorities forbade the staging of the play in Vienna until 1918, when it was performed for the first time six years after it had been written.

The question of interests here is whether Professor Bernhardi was right to insist on keeping the young woman in ignorance of her imminent death. Or was the priest right in wanting to administer last rites?

≥▲

There is a large body of literature dealing with what is considered the appropriate interaction with a dying patient. Although opinions by various authors diverge somewhat, modern medical professionals and ethicists in general opt for telling the truth: terminal patients should not be kept in ignorance.

The Department of Bioethics and Humanities of the University of Washington, for example, lists several questions an attending physician needs to consider, including whether the patient wants to know the truth and how much he or she should be told.[1] In the vast majority of cases, patients should not be kept in ignorance. Exceptions are acceptable if revealing too much information would have a harmful affect on the patient: for example, if disclosure would make a depressed patient actively suicidal. Different religious or cultural beliefs must also be considered. Many people with traditional Navajo beliefs, for example, do not want to hear about potential risks of treatment, because according to their beliefs to hear about such risks is to invite them to occur.

The *Archives of Internal Medicine*, a flagship journal of the American Medical Association, published the results of discussions with twenty focus groups of patients, family members, health care professionals, and physicians with expertise in end-of-life care. The authors concluded that "The top priority that came across . . . was the need for physicians to be honest and candid. . . . Any hope offered must be realistic and relevant. Physicians should be ready and willing to discuss dying with patients who are nearing the end of life, and with their families."[2]

Some doctors take their truth-telling too far. One critical care physician described her harrowing experience with a dying patient. As soon as she explained his prognosis, he lashed out, insisting there was nothing wrong with him. Obviously, he did not want to know the truth. Angered by his uncooperative behavior, she insisted, speaking loudly: "You're dying . . . it could be hours now. I don't think you will make it through the night." He did not.

The experience seems to have haunted the good doctor. In an apparent effort to unburden herself, she described it in an OpEd essay in the *New York Times*. "Denial was my patient's only defense mechanism. And as soon as the words left my mouth, I realized how cruel it was to try to take this defense from him in the final hours of his life. . . . I know that I added to my patient's pain in the last hours of his life. I wish that I had done it differently."[3]

More than one thousand three hundred readers' comments and letters poured in at the NYT's editorial office. Although nobody excused the way in which the doctor delivered the news to her patient, the overwhelming majority believed that she had done the right thing. A patient has the right to know about his state. But that is the answer to a different and trivial question. Of course, the patient has the right to know. The question here is: Does the patient have a *duty* to know?

Then the magazine *Psychology Today* weighed in on the issue. "Physicians who lie to dying patients may soothe their anguish but also erode trust in the profession," the author, a member of the ethics committee of a major hospital on Long Island, wrote. With this ambiguous statement, a clear answer is circumvented. Is it the doctor's duty to "soothe the patient's anguish" or to ensure "trust in the profession"? The oath of Hippocrates says "do no harm" and targets the patient, not society. The interaction at the deathbed, however, is between the doctor and the patient, not with society as a whole. The author's conclusion is no less enlightening: "For the sake of future patients and the integrity of the medical profession, doctors need to learn the skill of telling the truth without doing harm."[4]

Is the answer to the question—whether a dying patient should be told the truth—that a doctor should uphold the profession's integrity for the sake of future patients? Does a dying patient care about the trustworthiness of the medical profession after his or her death? Well, yes, a patient may want to uphold his children's and grandchildren's trust in the medical profession; but this same patient may wish to die peacefully without knowing of his or her own imminent demise.

As to the last rites for Catholic believers, the sad issue still crops up from time to time. On October 15, 2021, the British parliamentarian Sir David Amess, a practicing Catholic, was attacked and mortally wounded by an Islamist terrorist in a church where he was holding a townhall meeting with his constituents. Stabbed multiple times, he was close to death. A priest who had heard of the attack rushed to the church and asked a police officer if he could deliver the final sacraments to the dying man: "A Catholic when they're dying would want a priest there."

However, since this was an active crime scene, the police barred him from entering the church. So, instead of administering the last rites, the priest prayed the rosary outside the police cordon with a fellow parishioner. The parliamentarian died, and after the public's horror about the terror act had died down somewhat, the incident with the priest became a widely discussed issue.

In the aftermath, police guidelines in England and Wales were changed to allow clergy to help meet the religious needs of badly injured crime victims. In addition, formal legislation is being discussed in the parliament to ensure that Catholic priests are allowed to give the last rites to the dying, even at crime scenes.

So we are back to Professor Bernhardi's dilemma . . .

32

THE VEIL OF IGNORANCE

John Rawls's Theory of Justice

Imagine that you live in total freedom among a group of people unencumbered by traditions, customs, and any other restrictions. Would that be the pinnacle of joy? Maybe not so much. There would be no government, no police, no fire department, no traffic laws, no court of justice; life would be totally free but totally lawless. As the political philosopher Thomas Hobbes (1588–1679) wrote in his magnum opus *Leviathan*, there would be no culture, no navigation, no knowledge of the face of the earth, no arts, no letters, no society; instead, there would be rapes, thefts, murders, and continual fear of violence. Human life would be "solitary, poor, nasty, brutish and short."[1]

As more and more people began to live in close proximity, they realized the need for some sort of arrangement among themselves as an alternative to the chaotic state of nature. Such an arrangement, generally called a social contract, is based on the explicit or tacit consent of the citizens. In exchange for the curtailment of some freedoms, the people submit to an authority, thereby gaining protection and security. Only then is it possible to live in an orderly and safe manner within a community. Hence, to form a society, the individuals must enter into a contract that defines their moral and political obligations.

An early discussion about social contracts comes down to us in Plato's dialogue *Crito* in which Socrates, condemned to death for allegedly corrupting the youth, refused an offer to escape from prison. He explained to Crito that as a citizen of Athens he had agreed to live by the nation's rules, even if they seem unfair in his case. A state cannot exist without rules, and citizens are not allowed to decide whether to follow or disobey them. By residing as a citizen in Athens, Socrates implicitly agreed to abide by the nation's laws. From this it follows that he must accept the punishment that the nation may impose on him. That is the nature of the social contract.

Another notable thinker after the ancients who addressed this question was Thomas Hobbes. He agreed with Socrates and gave his reasoning a solid footing. Because people are self-interested but rational, it is in their interest to live together in a civil society. Hence they must be willing to give up some freedoms and agree on a social contract that encompasses laws and enforcement mechanisms. Even if the authority manages affairs poorly, be it a benevolent king, a tyrant, or a plodding government, the situation will be better than the original state of nature. Nobody may resist the authority's power to defend order and protect society from a chaotic state of nature.

But what form should a social contract take? On what principles of government could or should a community of people agree? Political philosophers have been wrestling with this question ever since Thomas Hobbes, John Locke (1632–1704), Jean-Jacques Rousseau (1712–1778), and John Stuart Mill (1806–1873) all weighed in on the subject.

One suggestion put forth by the English jurist and social reformer Jeremy Bentham (1748–1832) is called utilitarianism. It consists of measuring people's utility for the pleasures that governmental policies provide, minus the pains. And then it becomes the government's objective to provide the greatest happiness for the greatest number of citizens; the sum total of the citizenship's happiness should be maximized. Bentham even invented something that he called the felicific calculus, which would measure *hedons* for pleasures and *dolors* for pains.

Widening a sidewalk would provide pedestrians with additional utility while creating disutility for horse-drawn carts. According to Bentham, all one has to do is add the *hedons* of all pedestrians, deduct the *dolors* of all cart-drivers, and if the result is positive, then widen the sidewalk, otherwise don't. But the problem with utilitarianism is that just as the subjective, first-person experiences of sensory perceptions cannot be compared among individuals, neither can pleasures and pains (see chapter 30 on qualia). One person's dislike of beetroots cannot be compared by another person's love for tomatoes. Hence people's utilities cannot be summed in a meaningful way. But even if one could, would the policy be just? Would it be fair?

A century after Bentham, the Italian sociologist and economist, Wilfredo Pareto (1848–1923), developed his principle of economic welfare, known as Pareto optimality. Whereas Bentham strove to maximize total welfare, Pareto avoided utility comparisons by emphasizing efficiency. His notion of the best of all worlds was a state in which no one could be made better off without making someone else worse off. Regrettably, neither Bentham nor Pareto had any regard for fairness.

੨ఊ·

Enter John Rawls (1921–2002), one of the foremost philosophers of the twentieth century. His book, *A Theory of Justice*, published in 1971, sold hundreds of thousands of copies in the United States alone.[2] In his view, utilitarianism was defective

John Rawls

because "the greatest good of the greatest number" ignores what he called "the priority of the right over the good." Similarly, he criticized Pareto's notion of efficiency because it permits highly unequal distributions as long as no one is made worse off. In stark contrast, Rawls's principle of "justice as fairness" regarded fairness to be paramount, treating efficiency as a secondary concern. Whether policies resulted in advantages for a certain part of the society was less important.

The challenge now was to find a procedure that would produce a social contract that would be fair and acceptable to all. That is no easy task: the poor want welfare, the rich want tax breaks; men want to be the breadwinners, women want equal pay; the gifted want advancement based on merit, the untalented want remedial help; minorities want affirmative action, majorities complain about reverse discrimination. And adherents of each religion want their church to be the dominant one. To bring all wishes under one umbrella is impossible.

To show how a random group of people would come up with a just social contract, fair to all, Rawls devised a thought experiment. In the original state of nature, ground zero of social contract theory, a group of people would come together to decide on the principles that would permit them to live together in an orderly community. Rawls calls this the "original position." The crucial point in Rawls's thought experiment was that these people, when making their decision, should act behind what he called a "veil of ignorance": they are ignorant as to their ethnicity, race, gender, age, social status, talents, intelligence, and disabilities. Also, they do not know what type of life they would consider to be a good life. I will call these ignorant people proto-citizens. Because the conditions in the original state of nature are fair, whatever social contract these proto-citizens come up with will be just.

So what are the principles on which these proto-citizens, in their eukaryotic state, will agree? Being ignorant but rational, the proto-citizens will necessarily settle on what Rawls called the Two Principles of Justice. The first: every citizen must have as much liberty as possible as long as everybody else possesses the same liberties. The second: social and economic qualities must

(a) be available to all, and (b) be to the greatest benefit to the least-advantaged member of society.

The first principle, affording everybody equal basic rights and liberties, supersedes everything else and should be anchored in the nation's constitution. The second principle, regulating the distribution of wealth, position, social status, etc., says that inequalities are permissible and even desirable but must benefit the least advantaged. For example, when a talented hi-tech engineer becomes manager of a start-up company, the company creates jobs and thus will be of benefit to society. But the proto-citizen is ignorant as to whether she will turn out to be the hi-tech manager making a ton of money or the janitor living on a minimum wage. Because being rational means being risk averse, she will opt to maximize the minimum wage. In economic lingo, she will opt for the *maximin*.

Had she assumed that she would turn out to be the manager, she would have opted for *maximax*, for maximization of her compensation, by lowering the minimum wage. But that would have been risky: it could have turned out that she would become the janitor. And being rational, as Rawls stipulated that proto-citizens are, means being averse to risk. Hence the social contract on which proto-citizens situated behind the veil of ignorance would agree, arranges social and economic inequalities in such a manner as to make the worst off as well off as possible; economic policies would benefit the well-being of the least advantaged.

Rawls's two principles stand in contrast to modern versions of Bentham's utilitarianism, as set out for example by John Harsanyi (1920–2000), recipient of the 1994 Nobel Memorial Prize in Economic Sciences. According to his theory, rational, self-interested people in the original proto-state are not risk-averse but strive to maximize their expected utility. According to modern utilitarianism, this makes society better off and the social contract should reflect this. Hence, if managers' utilities for higher compensation more than offset the resulting disutilities of the minimum wage earners, then so be it. Life just ain't fair.

But if justice is to be understood as fairness, as per Rawls's scheme, ignorance is the key concept. Even when they eventually get to know their economic and social status in real life, nobody can complain about being treated unfairly because everyone explicitly agreed to the policies in the proto-state, before the veil of ignorance was lifted.

33

SHOW AND (DON'T) TELL

Zero Knowledge Proofs

I n the one hundred sixteenth episode of the sitcom *Seinfeld*, an odious character called the "Soup Nazi" sells soups based on secret recipes. Elaine gets ahold of his recipes and reads them aloud for everybody to hear. The secrets are out, and the Soup Nazi's business is ruined. (The plot is a bit more intricate . . . watch it.)

Elaine could have done better. By keeping the secret recipes to herself, she could have opened a rival soup stand. To customers, Elaine's soups would be the real thing, but only she and the Soup Nazi would know what was in them. By selling identical soups, Elaine would have proved that she actually knows the secret ingredients, without giving away the actual knowledge. This is one aspect of what is known in computer science as a *Zero Knowledge Proof*. A second aspect of Zero Knowledge Proofs is discussed later in this chapter.

There are many instances in life when one may want to show and not tell: for example, to prove a fact without actually divulging it. You may want to prove to a bank that you have a good credit score without actually disclosing it; you may want to convince an employer that you do not have a certain health condition without revealing your complete medical history; or you may want to show the bartender that you are of drinking age without revealing how old you are.

A pharmaceutical company, involved in a patent dispute, may want to prove to the judge and the jury that it knows the ingredients of a compound without revealing its composition. In disarmament talks, a nation may want to prove that it has nuclear capabilities without actually exploding an atom bomb.

≥⩗

An early example of a Zero Knowledge Proof was given by an Italian mathematician in the sixteenth century. Until then, it was received wisdom that no formula existed to solve cubic equations: $x^3 + ax^2 + bx + c = 0$. But then Niccolò Tartaglia (1500–1557) unexpectedly claimed that he had discovered the formula to solve cubics. However, to remain unrivaled, he would keep it a closely held secret. He was the sole person in the entire world who knew how to solve cubic equations, and he intended to keep it that way. A compatriot, Gerolamo Cardano (1501–1576), was very intrigued. He was working on a book about arithmetic and was about to claim that cubics could not be solved.

How could Tartaglia prove to Cardano that he knew the secret formula, but without divulging it? Easy: he told Cardano to send him a few dozen cubics, and by return mail sent him the solutions . . . albeit without letting on how he arrived at them. Cardano was baffled. He could easily verify the correctness of the answers by plugging them into the cubics but was left frustrated at not knowing how Tartaglia had done it. There's more to the story—accusations of heresy, plagiarism, double-crossing, and broken oaths—but we won't let ourselves be sidetracked here.[1]

Here's Tartaglia's equation that solves the cubic:

$$x = \sqrt[3]{\left(\frac{-b^3}{27a^3} + \frac{bc}{6a^2} - \frac{d}{2a}\right) + \sqrt{\left(\frac{-b^3}{27a^3} + \frac{bc}{6a^2} - \frac{d}{2a}\right)^2 + \left(\frac{c}{3a} - \frac{b^2}{9a^2}\right)^3}}$$

$$+ \sqrt[3]{\left(\frac{-b^3}{27a^3} + \frac{bc}{6a^2} - \frac{d}{2a}\right) - \sqrt{\left(\frac{-b^3}{27a^3} + \frac{bc}{6a^2} - \frac{d}{2a}\right)^2 + \left(\frac{c}{3a} - \frac{b^2}{9a^2}\right)^3}} - \frac{b}{3a}.$$

(where the x can become one, two, or three real or complex solutions).

❧

To show how Zero Knowledge Proofs work in practice, let the Soup Nazi challenge Elaine to a public contest. He provides her with bowls of soup, some of which are "Jerry's favorite," the others "Kramer's favorite." Elaine needs to show the Soup Nazi that she is able to distinguish which is which, just by looking at the bowls. It could be by color, by texture, by smell, or by extra-sensory perception. But she does not want to tell him how she knows. She also has a second objective (alluded to above): she wants to leave the onlookers in doubt. Only the Soup Nazi is to be convinced that she knows the secret (albeit without divulging it); the onlookers should know nothing at all.

The challenge starts by the Soup Nazi calling out "*Jerry's favorite!*" Elaine picks a bowl and gives him a taste; it is, in fact, Jerry's favorite. The Soup Nazi is not yet convinced; after all, there is a 50 percent probability that Elaine picked the right soup just by chance. So he tests her again: "*Kramer's favorite!*" Elaine picks a bowl and gives him a taste. Right again! The Soup Nazi is still not convinced: there is a 25 percent chance that Elaine was lucky twice in a row. So he challenges Elaine again and again and again, and every time that Elaine picks the right soup the evidence that she is able to distinguish the soups just by looking at them becomes stronger and stronger. Indeed, the probability of her picking the right bowl just by chance goes down by 50 percent every time. After the tenth exchange, the probability of Elaine having picked the correct bowl every time, purely by luck, is only one in more than a thousand. With continued exchanges, the probability of the Soup Nazi being wrongly convinced of Elaine's knowledge can be made arbitrarily small.

What about the onlookers? They simply saw exchanges between Elaine and the Soup Nazi, with the latter admitting every time that the soup was the correct one. But for all they

The actor Larry Thomas who played the Soup Nazi

know—even though, given the animosity between Elaine and the Soup Nazi, it may be difficult to believe—the whole thing could have been a set-up; the Soup Nazi and Elaine could be in cahoots; they could have prepared the series of exchanges in advance.

Although the Soup Nazi has been convinced (down to a minute probability of error) that Elaine is able to distinguish the two kinds of soup—even though he has zero knowledge how she does it— the gist of it all is that the onlookers have zero-zero knowledge: not only do they not know how Elaine does it, they don't even know whether she is at all able to do it.

34

LIMITED LIABILITY

The Corporate Veil

Gregory is a gifted engineer. He recently invented *Food-Pro*, a new kind of food processor. All he needs is one thousand dollars to build a prototype. He assembles nine friends, and he and they all invest one hundred dollars. The expectation is that, eventually, each one of them will get one tenth of whatever the profits will be. Should it turn out that Gregory's machine is unworkable, they lose their initial investment, one hundred dollars. But Gregory is lucky: he builds the machine, it works, and he sells it to Martha for $1,500. He and his friends pocket $150 each; they all made $50 in profit and everybody is happy.

But then, horror of horrors, the food processor blows up and destroys Martha's kitchen. The damage amounts to $2,000. Martha goes to court. Not only does she want her money back, but she also demands $2,000 to compensate her for the damages; altogether she sues for $3,500. Martha wins the case, and the ten friends must each pay her $350. After deducting the initial receipts ($150), they all lost not only their investment but also $200 on top of that.

Undeterred, Gregory goes back to the drawing board and designs *FoodPro Mark II*. He again hits on his friends, but this time he hits a wall. They'd be prepared to invest $100 again, but they are not willing to assume the risk of becoming responsible for more potential damages. So thanks, but no thanks! *FoodPro Mark II* stays on the drawing board and never gets built.

This is a shame. Gregory had some great ideas, and if he could have brought them to fruition, there would have been real progress in the food processing world. But it is not to be.

This is why nations created so-called limited liability companies—to encourage economic progress without scaring off investors.

ॐ

Chinese traders in the third millennium BCE were some of the first businesspersons to reduce the risk of doing business. Because ships were prone to maritime disasters, groups of traders pooled their cargo and distributed it among several vessels. In the event of a shipwreck, only a part of every trader's wares were lost. This was the first example of a company of shareholders, with each member holding a certain share of the combined enterprise. In the second millennium BCE, the Code of Hammurabi recognized partnerships; in the sixth century CE, Emperor Justinian allowed for formation of collective bodies; and in the Middle Ages, trade unions formed guilds.

The Renaissance and the Enlightenment period saw the creation of stock companies such as the English East India Company (1600), the Dutch East India Company (1602), and the Hudson's Bay Company in Canada (1670). Shareholders owned certain proportions of a business enterprise in accordance with the number of shares they held.

Insurance companies such as Lloyd's of London, for example, started operating in the late seventeenth century and were also such partnerships. So-called underwriters assessed the risks of

Coat of Arms, for Lloyd's of London

ships' voyages and collected the insurance premiums. If a boat and its cargo were lost at sea, the underwriters had to reimburse the traders for their wares and the shipowner for the boat. The underwriters were personally responsible, and their liability was unlimited. Whenever a maritime disaster occurred, the payouts surpassed the premium many, many times; it could reduce an underwriter to poverty. Similarly, in early times, stock companies could call on their investors to pay debts to creditors. So underwriting and investing were activities reserved for well-to-do individuals who could weather a disaster or two.

As economies progressed, it became necessary to enlarge the pool of people who would provide funds for new ventures. At the same time, the risk of doing business became even greater. Today customers who feel that they received deficient service or a faulty item can demand compensation not only for what they paid but also for collateral damage. Design faults, unclear instructions, mislabeling, unforeseen accidents, and inadvertent misuse can be grounds for lawsuits. Producers of goods and providers of services who did not uphold their respective side of an agreement to the fullest can be sued for reimbursement, damages, and even for emotional distress.

It is fair and reasonable, of course, that customers should be indemnified for faulty products. However, that creates a problem for the economy. For an economy to grow, funds must be provided. In a free market-based economy, these funds are provided by investors. Today, risk-averse investors—and except for bungee jumpers and skydivers all people should be risk averse—are not prepared to cover liabilities that surpass their investment. They are prepared to lose the funds they originally invested, but not more than that. If they can also be held accountable for millions of dollars in damages, they simply won't provide funds. And without these funds, the economy would stagnate.

Hence, to nudge the economy forward, investors must be shielded from liability. It is for this reason that the *corporate veil* was invented. This notion indicates that the shareholders, the owners of the corporation, remain hidden from public view. Customers who would like to hold someone liable for their losses remain ignorant of who is behind that veil. The best they can do is sue the corporation. If the corporation has no money to settle its debts or cover damages, it may declare bankruptcy, in which case creditors and plaintiffs often receive nothing. The investors, on the other hand, remain unscathed, except for losing their initial investment.

❧

The corporate veil was first adopted in the United Kingdom in 1855: the Limited Liability Act established the legal separation between a company and its owners. Half a century later the United States followed suit. The DuPont Company manufactured gunpowder, which had the unfortunate tendency to blow up from time to time. To protect itself from lawsuits, the DuPont family convinced the Delaware legislature in 1899 to enact the General Corporation Law, which would shield the family behind a corporate veil, thus protecting it from being subject to unlimited liability.

The invention of the corporate veil was such an important development that Butler Murray, the President of Columbia University, stated in 1911 that "the limited liability corporation is the greatest single discovery of modern times. . . . It substitutes cooperation on a large scale for individual, cut-throat, parochial competition."[1] Maybe that was a wee bit exaggerated because some people would argue that the invention of anesthesia, the telephone, and the electric lightbulb all have a more legitimate shot at the title. But it showed that economic and social progress could come not just from scientific breakthroughs but also from legal and institutional innovations.

Corporations whose owners hide behind the corporate veil append *Ltd.* (limited) to their name, or *LLC* (limited liability company), or *GmbH* in Germany (*Gesellschaft mit beschränkter Haftung*). French companies state the fact that customers are kept in ignorance in their own way. They append *SA* to the company's name, which says it all: *Société Anonyme*, the anonymous company.

Note, however, that if owners of a corporation engaged in fraud, negligence, or other misconduct, the corporate veil can be lifted: by order of a court, limited liability can be disregarded and the owners' identities divulged. In such cases, plaintiffs are no longer ignorant of the shareholders' identities and may become personally liable for the corporation's obligations.

35

MOVE TO STRIKE

Unringing the Bell

We all know the "*move to strike*" interjection by a prosecutor or by a defendant's attorney, if not from real courtrooms, then from television and movie trials. The interjection may be uttered because the attorney's question is leading, overly aggressive, lacks foundation, or because the witness's testimony is insufficient, immaterial, redundant, or based on hearsay.

If a judge believes that the question is, in fact, legitimate and the witness's answer is pertinent, then she will overrule the objection and direct the witness to answer the question. However, if a judge feels that the question does not conform to court procedure or the witness's answer is evasive, overly broad, or not to the point, she may sustain the objection. In that case, the judge having sustained the objection, instructs the jury to ignore the question and the witness's answer.

Well, there's a problem. The 1959 movie *Anatomy of a Murder*, one of the finest trial movies ever made, puts the matter clearly. The defense attorney, played by James Stewart, asks the witness to confirm that the prosecution bungled the investigation. Of course, the prosecutor immediately objects, and even before the judge can admonish him, Stewart withdraws the question.

James Stewart and Ben Gazzara in the courtroom drama

There is nothing left for the judge to do than to declare: "The question and answer will be stricken, and the jury will disregard both the question and the answer." As Stewart returns to his seat, Ben Gazzara, playing the murder suspect, whispers: "How can a jury disregard what it's already heard?" "They can't Lieutenant," Stewart whispers back, shaking his head, "they can't." What the astute defense attorney was saying is that it is impossible to "unring a bell."

This phrase was coined in a Supreme Court case in Oregon State in 1912. A man named Fritz Rader had been found guilty of arson committed four years earlier, on October 29, 1908, by burning two stacks of hay. Rader appealed, contending that in spite of objection by his lawyer, inadmissible evidence had been introduced during the trial. Namely, the judge had allowed the prosecutor to ask the plaintiff what had occurred on July 12, 1908, a date three months previous to the arson. The plaintiff testified that the defendant and a friend "passed by where I was at work in my field. They came through my place and went down by the house and cut one of my milk cow's tails off."

The judge realized that the mutilation of the cow, despicable as it was, had nothing to do with the arson and that the information was prejudicial to the defendant. He immediately instructed the jury to disregard the answer. Thus the error of inadmissible evidence was evidently rectified; the jury deliberated and declared Rader guilty of arson.

The Supreme Court saw things differently. The question was whether an express instruction to the jury to disregard testimony rectifies the error. Sometimes it does, sometimes it doesn't. "Although the withdrawal from the jury of evidence erroneously admitted may sometimes cure the error, such withdrawal must be so emphatic as to leave no doubt in the minds of the jurors as to the unequivocal repudiation by the court of such erroneous evidence." But this is difficult, if not impossible, because "even then it will not always be held that the error was cured." After all, it is hard to "remove from the mind an impression once firmly imprinted there," the Supreme Court wrote, famously adding that "it is not an easy task to unring a bell." Rader's judgment was reversed and he was granted a new trial.[1] Whether he was eventually convicted is not known to this writer.

୧ବ

To unring a bell, even in the figurative sense, is, indeed, impossible. In the mid-1950s, a contractor allegedly paid the mayor of Baxley, Georgia, sums of money in order to be awarded a contract. In the closing arguments, the prosecutor addressed the (all-male) jury thus: "Whether those amounts were reimbursement for expenses or kick backs—any of you gentlemen that know anything about politics, when you throw out that much money, why, somebody is going to have to pay somebody else."

Pouncing on this outrageous statement, the defense counsel immediately moved for a mistrial. In an attempt to mitigate the damage, the judge told the jury: "Just disabuse your minds of that statement, gentlemen, and don't let it influence you in any way. I am sure [the prosecutor] did not intend to say it, and he should not have said it, but just remove that from your mind in the trial of this case, and with that I overrule your motion [for a mistrial]."

The Court of Appeals, where the case subsequently landed, took a dim view of the handling of the situation by the lower court judge. To imply that all politicians are corrupt is obviously prejudicial and to simply ask the jurors to "disabuse their minds" from that statement is disingenuous. "One cannot unring a bell," the Court of Appeals commented, and then, for emphasis, coined an even more memorable analogy: "If you throw a skunk into the jury box, you can't instruct the jury not to smell it."[2] The case was remanded for a new trial.

❧

Introducing prejudicial information into a courtroom trial may not be the lawyer's fault. The question may have been proper but the witness is nonresponsive. Nonresponsiveness has two meanings: too little or too much. A witness may refuse to answer, be evasive, be hostile, be cagey, or lie. In such cases, the interrogating lawyer may well wish to dig further with follow-up questions. If the witness continues to prevaricate, jurors will draw the appropriate conclusions. More serious, on the other hand, is the case when a witness is too talkative, volunteers irrelevant or unnecessary information, rambles, strays from the question, or expands on it. Even though the judge may instruct the jury to ignore the answer, jurors will draw conclusions just the same because it's impossible to unring the bell. Picture the following exchange.

> PROSECUTOR: Mrs. White, please state your full name for the record.
>
> THE WITNESS: My name is Maggy White; the same Maggy White that dirty, rotten ex-con over there beat to a pulp, kicked when she was down, and spat on. I tell you that bum has been terrorizing my neighborhood for years, and if this jury doesn't have enough guts to send him up this time, I'm going to get my gun and take care of him myself.[3]

Even if the judge strikes the unresponsive answer from the record, it is difficult to imagine that the jurors will strike the information from their minds.

Of course, a "yes or no" question should be answered with a yes or a no. Adding a lengthy explanation is considered unresponsive. Sometimes, however, an unresponsive answer may be to the advantage of the interrogating lawyer, although the opposing lawyer may want to strike it. Consider the following (imagined) exchange:

> PROSECUTOR: Mrs. Brown, did you see what color the light was when the defendant's car entered the intersection?
>
> WITNESS: Yes, the light was red; but the defendant couldn't see it, because his view was blocked by the big whisky bottle he was drinking out of.

Clearly, the answers by Mrs. White and Mrs. Brown are unresponsive, and the defendants' counsels would love to object. But they cannot because they have no standing to do so. The requirement that witnesses be responsive is designed to protect the procedural rights of the interrogator, not of the opposing counsel. If unsolicited testimony helps the interrogator, then so be it.

≥•

The objective of a trial is to find justice. Procedures have been put in place so that the search for justice can proceed in an orderly and fair fashion. Jury trials are fraught with potential pitfalls, but by and large a panel of peers does judge fairly. One of the pitfalls, however, is the introduction of unwarranted information. The judge can instruct the jury to ignore it, and jurors may try their best to follow the instruction. But just as a bell cannot be unrung, or a skunk's stench be unsmelled, a prejudicial statement cannot always be "cured" by the judge's instruction to disregard it. On occasion, the resulting miscarriage of justice can only be remedied by appointing a fresh jury in a new trial.

36

HOT STOCK TIPS

Insider Trading

Marco, a well-to-do investor, gets a tip from his brother whose golf partner's wife is the office manager of PQR Corporation's CEO. The corporation, she told her husband, who told his golf partner, who told his brother, was about to close on a particularly lucrative contract. Marco immediately called his broker and told him to buy a million dollars worth of shares in PQR Corporation. Several weeks later, the lucrative contract is made public. Overnight, PQR's shares rise by 50 percent: Marco sells his shares, makes half a million dollars in profits . . . and goes to jail for two years. He was lucky; he could have received a sentence of twenty years, more than for murder in some states.

Suzy overhears a conversation between two clients at the hairdressers. They are sales managers at ABC Corp. and discuss a big lawsuit that one of the corporation's customers would be filing against ABC in ten days. They know that ABC does not have a leg to stand on in court and that the share price would drop precipitously as soon as the news hits the wires.

If Suzy had owned shares, she would now have been tempted to sell them off quickly, before the news becomes public knowledge and the price drops. Luckily, she has none and, therefore, is in no danger of being lured into an illegal insider trade. But is there a way she could profit from what she overheard? Can knowledge about the calamity that is about to befall ABC Corp. benefit someone who does not own any shares in the corporation?

In principle, yes. Suzy could enter into an agreement with someone who wants to buy ABC shares at today's price but, in exchange for a discount, is willing to wait two weeks for delivery. Since the date of delivery will be after the news has become public, the price will have dropped by then. Suzy would then buy the shares on the stock market at the lower price and deliver them to the buyer, keeping the difference for herself. In financial lingo, this is called a short sale. And, guess what? It is also forbidden. Short sales based on information overheard at the hairdressers are illegal.

❧

The offense for which Marco and Suzy could be prosecuted is known as "insider trading." Actually, this designation is a misnomer. First of all, neither Marco nor Suzy were insiders. It was Marco's brother's golf partner's wife and the ladies at the hairdressers who were the insiders. Second, trading as such in the corporation's stock by insiders is not forbidden. Marco's brother's golf partner's wife is allowed to buy and sell the corporation's shares, and the ladies at the hairdressers are permitted to short sell the shares. It is only when trades take place on the basis of "material nonpublic information," aka insider knowledge, that they are illegal.

It may be very frustrating to investors, but whenever someone has information about a publicly traded company that has not been made known to the public, he or she is not allowed to act on it. Only people ignorant of material nonpublic information are allowed to trade. Insiders must disclose the information before they trade, or they must abstain from trading.

The first lawsuit against an insider was decided in 1909 when the US Supreme Court ruled that a corporate executive was guilty of fraud for buying a large number of shares of company stock without disclosing that he knew that the stock was going to jump in price.

In 1933, four years after the Wall Street Crash of 1929, Congress enacted the Securities Act, and a year later established the Securities and Exchange Commission (SEC), the government agency responsible for establishing, overseeing, and enforcing laws pertaining to securities fraud. At the time, the act prohibited fraudulent sales but said nothing about fraudulent purchases. It was Rule 10b-5, enacted in 1942, that closed that loophole. It states that it is illegal for any person to defraud or deceive a trading partner, including through the misrepresentation of material information, with respect to "the sale or purchase" of a security. In other words, the rule covers all instances when someone uses nonpublic information to profit from stock trades.

Why is Rule 10b-5 important? Why are transgressions penalized as harshly as murder?

The existence of the stock market is of supreme importance for the functioning of an economy. Only if people are willing to invest their money can corporations raise funds and can the economy remain healthy and grow. But much depends on trust; the stock market is a fragile instrument and could be abused by dishonest players. If ordinary citizens believe that only individuals with insider knowledge will profit from stock market trades, the flow of money would dry up and the economy would collapse. Hence, to be healthy, the securities market must be perceived as fair. It is Rule 10b-5 that creates a level playing field.

Securities and Exchange Commission seal

But the burden of proof rests with the SEC, and neither prosecution nor enforcement are easy. In 1984, for example, a football coach attended a track meeting in Norman, Oklahoma. While resting on a deck chair, he overheard the CEO of Texas International Company (TIC), who was sitting several seats in front of him, mention the possible liquidation of Phoenix, a company owned by TIC. Since the value of Phoenix's assets exceeded the market price of its stock, the announcement of liquidation would raise the price of Phoenix's stock. The football coach shared the hot tip with some of his friends; they purchased Phoenix's stock and resold it at a substantial profit several days later.

The court found the CEO not guilty because "only when a disclosure is made for an "improper purpose" will such a "tip" constitute a breach of an insider's duty." The discussion with his wife did not amount to "improper purpose" at all.[1]

Surprisingly, the football coach and his friends weren't guilty either because "It must be shown that the tippee knew or should have known that there had been a breach by the insider," the court ruled. Hence "tippee liability cannot result from [the CEO's] inadvertent disclosure to [the football coach]" because "these defendants did not know, nor did they have reason to know, that the information they received was material, nonpublic information disseminated by a corporate insider for an improper purpose."

≥●

Even though insiders use shifty and ingenious ways to cover their traces, the SEC does have numerous successes. Consider the indictment of two mental health professionals, one in New York and another in California. In the first case, the psychiatrist, Dr. Robert Howard Willis, was treating the wife of the well-known banker Sanford Weill who was at the time lobbying to be named chief executive of BankAmerica. His appointment would have come with an infusion of one billion dollars capital into BankAmerica, so his appointment would have raised BankAmerica's share price. During the following two weeks, Dr. Willis made twenty-three purchases for a total of thirteen thousand shares of BankAmerica Corporation. After the public announcement of Weill's bid, Dr. Willis sold his shares at a moderate profit of about $27,000. (Weill was never named CEO of BankAmerica.)[2]

The judge did not hold back in his castigation of the psychiatrist: Dr. Willis had not only blatantly infringed on the laws concerning insider trading but had also violated the two-and-a-half-thousand-year-old Oath of Hippocrates that states: "whatsoever things I see or hear . . . in my attendance on the sick . . . I will keep silence thereon, counting such things to be as sacred secrets." Eventually, Dr. Willis was sentenced to five years' probation and a $150,000 fine.

The psychotherapist Mervyn Cooper in Santa Monica should also have let himself be guided by the Hippocratic Oath. A marriage counselor, his clients were a senior executive of the Lockheed Corporation and his wife. In the course of their therapy sessions, the executive mentioned that Lockheed was negotiating a major financial deal. Just hours later, the marriage counsellor purchased options and shares in Lockheed. When the firm announced its ten billion dollar merger with Martin Marietta a few days later, the price of the shares soared, making Cooper a profit of $53,000. The Securities and Exchange Commission brought charges against Cooper, which he agreed to settle for $110,000.[3]

The psychiatrist and the psychotherapist should both have remained ignorant.

37

CARNAL KNOWLEDGE

Adam and Eve

And the Lord God commanded the man, saying: "Of every tree of the garden thou mayest freely eat; but of the tree of knowledge of good and evil, thou shalt not eat of it . . ."

Genesis 2, 16–17

Bible expositors have been puzzling for centuries about the meaning and deeper implications of God's prohibition. Why should human beings not acquire the ability to distinguish between good and evil? After all, is it not the most basic principle of all religions that individuals should do good and eschew evil? How could they do that, if they are not aware of the difference?

Well, there's more to the story than simply ignorance of moral precepts. After being tempted by the snake and eating the forbidden fruit, "the eyes of them both were opened, and they knew that they were naked." All of a sudden they recognized their physical differences. So Adam and Eve did what any embarrassed person would do in that situation: they hid between the trees. When asked where he was, Adam answered,

Adam and Eve by Albrecht Dürer

"I was afraid, because I was naked; and I hid myself," whereupon God retorted, "Who told thee that thou wast naked?"

After some finger-pointing from Adam to Eve, and from Eve to the snake, God meted out punishments: to the snake ("upon thy belly shalt thou go"), to Eve ("in pain shalt thou bring forth children"), and to Adam ("in the sweat of thy brow shalt thou eat bread").

Partaking of the fruit of the tree of knowledge made Adam and Eve realize that they were naked, but is nakedness an evil? Hardly. Adam and Eve hid out of shame for having misbehaved; they were punished not for being naked but for the transgression. The punishments were retribution for scorning God's prohibition. The offense derived from Adam and Eve's desire for independence and autonomy. Eating the fruit symbolized the loss of innocence and the introduction of guilt and shame.

Why did God issue the injunction in the first place? A hint was given by the serpent when it told Eve that after eating the fruit "Your eyes shall be opened, and ye shall be as God, knowing good and evil." And, indeed, a few verses further down the Lord says—to no one in particular—"Behold, the man is become as one of us, to know good and evil." Apparently, God, in a desire to keep humans innocent, preferred man (and woman) to remain innocent and ignorant; knowledge of good and evil was to remain beyond human understanding.

There's another facet to the story. The verb "to know" in the Bible indicates more than just the acquisition of cerebral knowledge. After being expelled from the Garden of Eden, so it is written, Adam "knew" Eve. We understand what is meant because Genesis goes on to explain that, after having been known by Adam, Eve conceived and bore the son Cain. And a generation later, Cain "knew" his wife who bore him the son Enoch. In short, knowledge in the biblical context indicates sexual intercourse or, in other words, "carnal knowledge."

❧

The term carnal knowledge (from Latin *carnis* = of the flesh) suggests that corporeal knowledge of oneself and another human being can be acquired in the intimacy of the bedroom. But the qualifier "carnal" indicates that the familiarity thus gained is merely physical: to achieve an intellectual or spiritual understanding between two people, another type of knowledge is required.

In medieval canon law, carnal knowledge became the official euphemism for sexual misconduct, not only for rape but also for prostitution, extramarital affairs, and any other sexual vice. As can be gathered from the London Metropolitan Archives, for example, in February 1549, the Court of Aldermen in London sentenced to the pillory and to prison a butcher who had not only "detestably kept and carnally knowen and used" the wife of a fellow butcher living opposite, "dyrectly ageinst Godes lawe" but also—almost an afterthought—hired men to kill the husband.[1]

৯৮

The biblical sense of knowledge as carnal knowledge was picked up by no less an authority than William Shakespeare. In his 1604 play *Measure for Measure*, Angelo, who had spurned his ex-fiancée Mariana because her dowry had been lost at sea, wants to have sex with Isabella, a beautiful but devout nun. The idea came up—I won't go into all the details here—to let Mariana go to bed with Angelo, while making him believe that he would be having intercourse with Isabella. So, in total darkness Mariana takes Isabella's place, and Angelo unwittingly consummates his overdue marriage to Mariana. Angelo never catches on, continuing to believe that it was Isabella with whom he had been copulating. This leads to Mariana's remarkable line:

> I have known my husband; yet my husband
>   Knows not that ever he knew me.

To make sure that the audience understands what Mariana means when she says *knows*, she reiterates that Angelo is her husband:

> Who thinks he knows that he ne'er knew my body,
> But knows he thinks that he knows Isabel's.

To summarize: Angelo thinks he *knows* Isabella's body and *ignores* Mariana's, but *ignores* the fact that he actually *knows* Mariana's yet *ignores* Isabella's.

ë

The story of the Garden of Eden can be read as a cautionary tale about the dangers of uncontrolled or inappropriate sexual behavior, but it is equally a narrative about the peril of losing the innocence that comes with ignorance.

As a postscript, let me note that the story about old police reports abbreviating the phrase "arrested for unlawful carnal knowledge" as "arrested f.u.c.k." is most probably false.

38

IGNORE IRRELEVANT ALTERNATIVES

The Theory of Games

When pondering a decision, one considers several options. Some are irrelevant and could, nay should, simply be ignored. Often they are not, however, and that may lead to immense consequences. As an instructive example, consider the results of the US presidential elections in the year 2000.

Although Al Gore won the popular vote, the all-important electorate of the State of Florida went to George Bush, who obtained 2,912,790 votes. Gore, who got only 537 fewer votes, lost. Note, however, that Ralph Nader, the Green candidate obtained 97,488 votes. With just 1.6 percent of the votes, Nader had zero chance of winning the presidency. As an alternative to Bush and Gore, he was totally irrelevant; his candidacy should have been ignored. But it was this Green candidate's participation in the election that prevented Gore, the consummate environmentalist who received the Nobel Peace Prize and an Oscar for his efforts to save the environment, from becoming president. This momentous event shows what may happen when people neglect to ignore an irrelevant alternative.

෫෧

The writing had been on the wall for at least half a century. In 1952, the French economist Maurice Allais had organized a conference in Paris under the heading *Foundations and Applications of the Theory of Risk Bearing*. (Maurice Allais would win the Nobel Prize in economics in 1988.) He claimed that even rational people do not behave according to what he dismissively called the "American School of Rational Decision Theory," namely, the Theory of Games, developed by John von Neumann and Oskar Morgenstern several years earlier. In particular, Allais was skeptical about game theory's stipulation that rational people ignore irrelevant alternatives.

To make his point, he devised an experiment.[1] He asked participants at the conference, all well-versed in the theory of probability and hence very rational, to answer two questions.

Maurice Allais

(I encourage readers to answer the questions that follow themselves, before reading on.)

Question 1: Which of the following situations do you prefer?

(a) $1 million for sure,

               or

(b) 10 percent chance of receiving $5 million,

    89 percent chance of getting $1 million, and a

    1 percent chance of getting nothing.

Most of Allais's subjects answered that they preferred (a) over (b), the $1 million for sure rather than the gamble. Then Allais asked the second question:

Question 2: Which lottery do you prefer:

(c) 11 percent chance of getting $1 million, and

    89 percent chance of getting nothing,

               or

(d) 10 percent chance of getting $5 million, and

    90 percent chance of getting nothing.

Most of the respondents preferred the slightly lower probability of obtaining a substantially higher prize—they preferred (d) over (c).

As we shall now see, the answers to the two questions contradicted themselves, which means that even these sophisticated interviewees were irrational. We reformulate Allais's first question, noting that "for sure" is equal to 89 percent plus 11 percent, Hence, the first choice is between

(a') 11 percent chance of getting $1 million,

    89 percent chance of getting $1 million,

        or

    (b′) 10 percent chance of getting $5 million,

        89 percent chance of getting $1 million,

        1 percent chance of getting nothing.

Because "89 percent chance of getting $1 million" is common to both lotteries, this part of the lottery is irrelevant and should be ignored. Hence, the question Allais asked boils down to

    (a″) 11 percent chance of getting $1 million,

               or

    (b″) 10 percent chance of getting $5 million,

        1 percent chance of getting nothing.

So the preference by most of Allais's interviewees of (a) over (b) is equivalent to preferring (a″) to (b″). The choice can be explained by most people's dislike of risk.

    Now, let's turn to (c) versus (d). Noting that 90 percent equals 89 percent plus 1 percent, we rewrite these as

    (c′) 11 percent chance of getting $1 million, and

        89 percent chance of getting nothing,

               or

    (d′) 10 percent chance of getting $5 million, and

        89 percent chance of getting nothing, and

        1 percent chance of getting nothing.

Similar to above, "89 percent chance of getting nothing" is common to both lotteries, hence this part of the lottery is irrelevant and should be ignored. Therefore, the question Allais asked boils down to

    (c″) 11 percent chance of getting $1 million,

               or

(d″) 10 percent chance of getting $5 million, and

1 percent chance of getting nothing.

So, the preference by most of Allais's interviewees of (d) over (c) is equivalent to preferring (d″) to (c″).

And now comes the punchline! Note that (a″) is identical to (c″), and (b″) is identical to (d″). Nevertheless, very many people prefer (a″) to (b″), and at the same time prefer (d″) to (c″). What a paradox! By adding "89 percent chance of getting $1 million" to both lotteries, many people prefer (a) over (b). By adding "89 percent chance of getting nothing" to both lotteries, they prefer (d) over (c).

The crucial point is that most people do not ignore the addition of an irrelevant alternative. Once an irrelevant alternative is added, decisions may be reversed, just like the addition of Ralph Nader to the slate of candidates changed the US electorate's decision from an environmentalist Democrat to a conservative Republican.

❧

In 1961, Daniel Ellsberg, of later Pentagon Papers fame, concocted an experiment similar to the one by Allais that proved an even stronger point: many people behave irrationally, even given their own assessment of probabilities (see chapter 11).

39

BLIND AND DOUBLE BLIND
The Placebo Effect

Placebos—sugar pills, saline injections, syrup—that have absolutely no medically proven therapeutic effect are administered to people for two different reasons. One is to make use of the so-called placebo effect in order to relieve pain and to "cure" a patient from disease (the reason for the quotation marks will become clear in a little while); the other reason for the use of placebos is to perform medical and other scientific research. In the first case, only the recipients of the placebo are kept in ignorance of the noncurative effect of the "medicine"; in scientific and medical research, the investigators are also kept in the dark.

Derived from the Latin verb *placere*, meaning to please, placebos do exactly that: they please the patient without doing anything else to treat the illness. (If a patient anticipates negative side-effects of a treatment, and then does indeed experience negative symptoms even though the substance was inactive, the effect is called a nocebo effect [from the Latin verb *nocere*, meaning to harm].) Many clinical studies have confirmed that a placebo may, indeed, lead to a beneficial reaction. This is known as the placebo effect. The effect is caused by nothing more than the person's belief that a treatment will be effective. By the power of suggestion, the brain tells the body to release some chemical that relieves pain.

How placebos work is still the subject of ongoing research. Although the placebo does not cure the patient, it is thought that the mere belief that a substance will work activates the prefrontal cortex, the area of the brain that is responsible for high-level thinking. Thus the mere belief that one receives a treatment may provoke thoughts in the brain that, in turn, trigger signals to other brain regions that are responsible for the production of feel-good neurotransmitters such as endorphin and dopamine, or the love-hormone oxytocin. These chemicals have a strong calming effect, muffle perception of pain, and boost feelings of pleasure.

Although a placebo will reduce the symptoms that are controlled by the brain, such as the perception of pain, insomnia, emotional reactions, and mood disorders, it will not cure a disease. Furthermore, for placebos to have a beneficial effect at all, patients must believe that they receive real medicine. So here's the rub: placebos involve deception. Hence the question: Is it ethical to deceive patients, even if doctors are convinced that it is for the patient's own good?

If one takes the Hippocratic Oath as a guiding principle, there seems to be no problem. The oath says "I will do no harm" and "I will not give a lethal drug to anyone." Neither of the two instructions is violated when a doctor administers placebos. However, the Hippocratic Oath also says: "Into whatever homes I go, I will enter them for the benefit of the sick, avoiding any voluntary act of impropriety or corruption." This clause implies a relationship of trust between patient and doctor. By prescribing pills the doctor knows have no therapeutic value, that trust may be violated, even though it is meant for the benefit of the sick.

That supremely rational people with high IQs can inadvertently be soothed by placebo thinking may be illustrated by the following anecdote. It is said that when Albert Einstein came one day to the house of Niels Bohr, an equally renowned physicist, he noted a horseshoe, believed by superstitious people to fend off bad luck, nailed to the wall above the front door. The following exchange ensued:

EINSTEIN: "Herr Bohr, you don't believe in such nonsense, do you?"
BOHR: "Of course not, Herr Einstein! But I have been told that it works even if you don't."

❧

I now turn to the second reason for using placebos. To assess whether a medical substance is effective at treating a disease, a large number of participants are enlisted in a study. Half of them will receive the substance that is being tested (the experimental group), the other half are given sugar pills or saline injections (the control group). We already know that the fact alone of administering a placebo can lead to improvement in a patient's condition, and some sugar-pill recipients may get better due to the placebo effect. To counter this effect, the subjects are not told which pill they received. But the placebo effect also operates in the experimental group: some of the patients in the experimental group will also get better even if the substance is not effective. Because it may be assumed that the placebo effect will occur in both the sugar-pill group and the ineffective-substance group in about equal measure, the substance can be deemed effective only if significantly more subjects in the experimental group recover than in the control group.

But there is another reason to be wary of experiments using placebos. In principle, it should suffice that the participants in the study remain ignorant about which pill they received. It is natural, however, that researchers want their experiment to succeed, for reasons of personal fame, and all the more so if the substance that is being tested is designed to heal ill people from their disease. What could happen is that the scientists conducting the experiment could influence the results of the study, knowingly or subconsciously, by giving the members of the experimental group superior care, better treatment, more encouragement, or personal attention. This could increase the chances of recuperation, or even if the compound is ineffective it could induce a placebo effect in these favored patients.

To counteract such distortions, scientists perform so-called double-blind studies: everybody is kept in ignorance, neither the participants nor the researchers know who got what. This ensures that the results are not influenced by any preconceived notions or expectations held by the researchers or by the participants. The participants have only been informed that they either receive a medicine or a placebo.

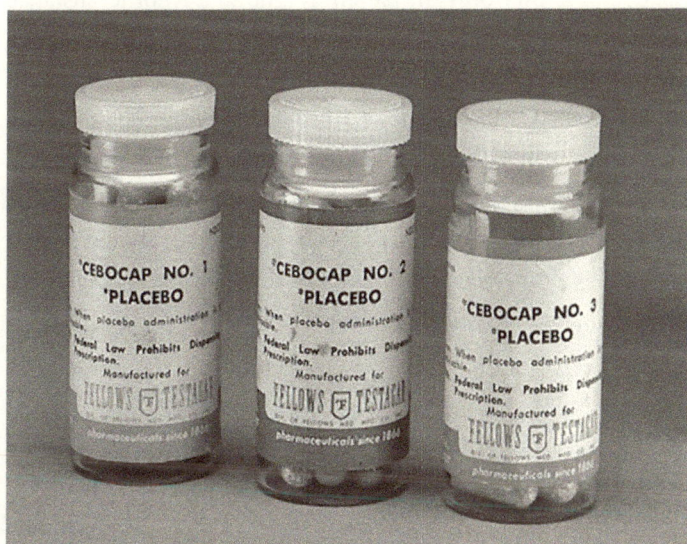

Placebos

Once the experimental stage has ended, and only then, are the results unblinded, and it is verified whether members of the experimental group showed a significantly greater success rate than the members of the control group.

It may seem a statistical unseemliness that during the first COVID-19 wave many people had to die before the vaccines of the pharmaceutical companies Pfizer and Moderna could be given the green light by the FDA. Only after a sufficient number of people had succumbed to the malady could the results of the studies be unblinded and it became possible to verify that the majority of the people who died had been administered the placebo.

ꙮ

By the way, are there also single-blind studies? The administration of placebos to relieve pain could be said to be single-blind: the doctor knows, only the patient is ignorant. Another instance of a single-blind study is, for example, the audition for a spot in a symphony orchestra. The candidate would play the instrument behind a veil. He or she knows, of course, whether he or she is a he or a she, white or colored, old or young. But the members of the jury remain ignorant of all this and will make their choice without bias. And for hiring decisions, candidates should submit CVs without any clues or hints as to their sex, race, religion, or any other characteristics. Employers should remain ignorant.

40

POLLS ARE LIKE PERFUME

The Bandwagon Effect

Public surveys have been conducted for about two centuries, mainly for elections but also for more mundane reasons. Corporations want to determine women's preferred beauty products, men's ideal shaving cream, young people's favorite pizza brand; music producers want to discover the best-loved artist; governments want to keep abreast of the public's feelings about their policies and about the satisfaction of their performance.

Most important, surveys are conducted before elections. Opinion polls are the backbone of electoral campaigns. Contenders commission surveys to assess their chances, candidates to tailor their messages, and the media to inform the public. By asking a sample of voters whom they favor, pollsters try to predict who will win the election. They often get close to what turns out to be the true result, but examples of polling gone wrong are legion. The foul-up of the 1936 presidential election—when the *Literary Digest* predicted an overwhelming victory for the Republican Alf Landon but the Democrat Franklin Roosevelt won by a landslide—is just one of many cases. A timelier example is the win of Donald Trump of the 2024 presidential election: in spite of polls predicting a razor-thin result one way or the other, he won all seven key battleground states.

But blunders are not the subject matter of this book; our concern lies elsewhere. Namely, polling is a two-way street: it not only evaluates the views and opinions of the public, it also shapes them. In a paper published in 1954, the later Nobel Prize winner Herbert Simon wrote that "voting behavior of at least some persons is a function of their expectations of the election outcome; published poll data are assumed to influence these expectations, hence to affect the voting behavior of these persons."[1] Many voters do not let themselves be influenced by the polls, but some people are more likely to vote for a candidate who is ahead in the polls, and a few others may opt for the candidate who is behind. The former behavior is called the "bandwagon effect," and the latter is the "underdog effect." There is also the "boomerang effect," which says that voters might remain at home instead of going to vote because one candidate is so far ahead in the polls that she is going to win anyway.

The bandwagon effect is the most common occurrence; it is similar to what is known as "herding behavior" in economics. If everybody invests in this or that market, then given the alleged wisdom of the masses I'd better do so too. In voting behavior, herding behavior may derive from the desire to fit in, to be on the winning side, or to avoid the effort of having to choose. Out of intellectual laziness—psychologists more politely call this "cognitive bias"—people use shortcuts and tend to "jump on the bandwagon" of the expected winner of an upcoming election. To determine who the likely winner will be, they peruse the polls.

Others vote strategically. Socialists could, for example, vote for the third-ranked Communist Party, in the hope that their preferred but second-ranked Socialist Party will form a coalition with the communists, thus stifling the chances that the front-running Conservative Party would form the government. And what do these socialists do to determine the ranking? They use the polling results.

And this is the problem: opinion polls can have a powerful influence, especially on undecided voters who very often are the ones who determine the outcome of close elections. (Polls can also influence the candidates themselves: based on surveys, potential candidates may decide to enter a race or to drop out of it.)

Hence, according to the bandwagon effect, last-minute publication of an opinion poll may sway undecided and hesitating voters in the direction of whoever is the front-runner. The poll results may be biased, prejudiced, random, erroneous, even fraudulent; the bandwagon effect may nevertheless determine the outcome.

How can the election authorities prevent such distortion? Easy! Simply keep the electorate ignorant of the results of voter surveys in the period before votes are to be cast, until closing of the ballot stations. As a result, there will be no bandwagon effect (and no underdog, no boomerang, no strategic voting), cognitive biases will be reduced, and nobody will be influenced by erroneous and falsified survey results. The expectation is that voters, kept in ignorance of opinion polls, will then elect whomever they deem most fit for the position.

To prevent distortions in the electoral process, many countries enacted laws that forbid the publication of surveys before an election. Many democracies, for example, France, Italy, Greece, Norway, and Israel, have blackout periods for poll results. Other democracies, including the United States, the United Kingdom, Germany, and Belgium, don't. In a survey of 133 nations, one-third reported blackout periods of one to six days, and one-fourth reported blackout periods of a week or more. However, one-third reported no blackout period.[2]

From a constitutional point of view, blackout periods are a problem. The prohibition of the publication of opinion polls impinges on freedom of the press, free speech, and the public's right to know. Thus countries are faced with a paradox: on one

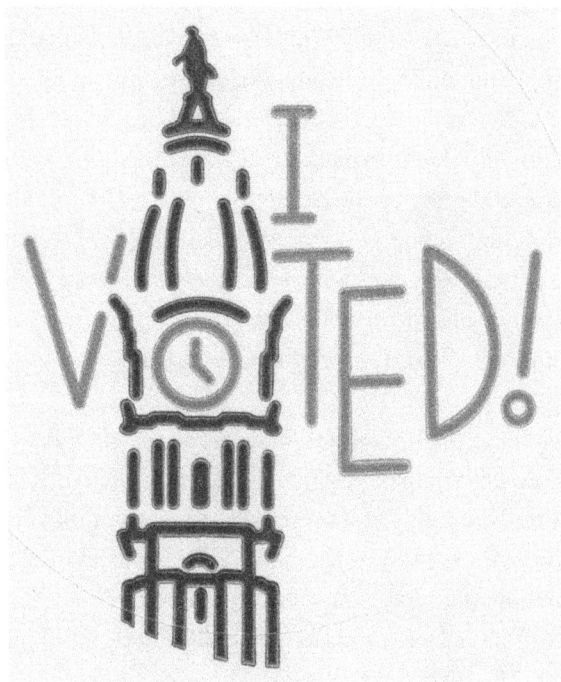

I Voted: a Philadelphia, Pennsylvania, sticker

hand, the right to know is paramount in a democracy; on the other hand, knowing the results of opinion polls distorts the voting process and hence impinges on democracy itself.

Few would condone Plato's view that ignorant masses should be prevented from ruling (and voting) altogether, but a good case can be made for keeping people ignorant of poll results. Countries have to decide for themselves which way to go. In any case, it is good to recall the advice that Shimon Peres, doyen of Israeli diplomacy, had for the public and for candidates: "Polls are like perfume, good to smell, poisonous to drink."

41

DON'T TALK ABOUT MONEY

Compensation Secrecy

You agree to follow the Company's strict policy that employees must not disclose, either directly or indirectly, any information regarding salary, bonuses, or stock purchase or option allocations.

The legality of clauses such as this one is questionable, and their enforcement is largely impossible. Nevertheless, they are routine in many employment contracts. The question is: Why are provisions about compensation secrecy often included in work agreements? Who, if anybody, is served by this secrecy? Who, if anybody, is hurt by this imposed ignorance?

The most immediately affected are employers and the employees. For both, salary secrecy elicits benefits as well as costs. (Further down on the list of those affected are the employee's colleagues, the employer's competitors, and the economy in general.)

For employers, the benefits of keeping employees ignorant about everybody else's salaries and perks is a paternalistic way of controlling the workforce. As long as employees do not know whether they are paid a competitive wage, compatible with their skills and abilities, they will be hesitant to take the risk of leaving their current position. The decreased labor mobility reduces the costs to the employer of hiring new workers to replace the ones who left for better-paying positions. Secrecy—a policy often camouflaged as "confidentiality"—also gives management the freedom of paying more to preferred employees and less to disfavored employees, without eliciting jealousy. Hence, somewhat paradoxically, pay secrecy may even support fairness in the workplace by allowing employers to set pay based on individual performance and qualifications rather than on factors such as seniority or what others are making. The biggest reason to keep salaries confidential is to mask pay differences between those performing the same job. Without pay secrecy, jealous workers would observe wage differentials without understanding the reasons, if there are any. Thus, to avoid conflict, employers try to compel their employees to keep their compensation secret.

There are more sinister reasons for corporations to keep information about compensation confidential. Pay secrecy allows immoral corporations to discriminate against the female workforce, minorities, and other unloved personnel. Pay secrecy can be abused because it prevents the detection of unfairness in corporations where inequities do exist, blocks claims of discrimination, and avoids lawsuits. Unfairness, inequality, and gender discrimination, even if only suspected, can create an unhappy work environment, but it is the very secrecy that may lead to discontent and a lack of trust among employees.

For employees, too, salary secrecy has its pros and cons. Foremost among the pros is the quest for privacy. Actually, this is odd because many people are willing to talk about politics, religion, their health, love, even sex—but salary? Never! Good reasons certainly exist for secrecy: high pay may be kept secret to prevent jealousy, low pay may be kept secret out of shame. Some employees want to protect a sweet deal that could be rescinded if it became public knowledge.

German pay slip, 1923

The cons are that employees remain ignorant about whether they are treated fairly and whether they are paid what they are worth. And privacy's adversary, the all too human trait of curiosity, remains unsatisfied. Finally, asymmetric information—as when the employer knows everybody's salary, but the employees don't—weakens the employees' negotiating position.

When weighing the costs and benefits of pay secrecy, for employers on the one side and employees on the other, no clear picture arises. Keeping salaries secret may simply be a question of social norms. By unstated convention, the discussion of monetary issues is generally frowned upon, and the flaunting of wealth in particular is taboo.

For the economy as a whole, pay secrecy is detrimental. To have an efficient labor market—with workers migrating toward positions where they are most valued, and corporations hiring the most suitable workers—all information about wages is indispensable. Asymmetric information due to compensation secrecy leads to inefficient allocation of human capital in society.

Employees, ignorant about where they stand in relation to others, lose out on better paying positions, and corporations and organizations do not obtain the best employees for the jobs. In general, pay secrecy contributes to income inequality and social injustice; society suffers and the economy as a whole is harmed.

≥⁂

To prevent discrimination and injustice, the law in the United States specifies that clauses in employment contracts about pay secrecy cannot be enforced. The National Labor Relations Act, first enacted in 1935, prohibited virtually all private-sector employers from limiting employees' activities related to "collective bargaining or other mutual aid or protection." This means that commercial companies cannot forbid discussions among their employees about their wages. The prohibition was extended to federal employees and contractors in 2014 by President Obama. His Executive Order specified that "When employees are prohibited from inquiring about, disclosing, or discussing their compensation . . . compensation discrimination is much more difficult to discover . . . and more likely to persist."

42

CURSED BE HE

Spinoza's Theses

Cursed be he by day and cursed be he by night; cursed be he when he lies down and cursed be he when he rises up. Cursed be he when he goes out and cursed be he when he comes in . . . no one should communicate with him, not even in writing, nor accord him any favor nor stay with him under the same roof nor [come] within four cubits in his vicinity; nor shall he read any treatise composed or written by him.

Who was it against whom this condemnation, harsh even by the standard of the time, was announced by the orthodox rabbis in Amsterdam in the year 1656? What communications and writings were to be utterly ignored by the Jewish public?

The excommunicated sinner, a twenty-three-year-old Dutchman of Portuguese-Jewish origin by the name of Baruch de Spinoza, is today considered one of the most important thinkers of the early Enlightenment. The theses that he proclaimed even at this early age so infuriated the Jewish community of Amsterdam that the elders expelled him from their midst. Henceforth he made his living as a lens grinder. After his death, at age forty-four, the local authorities forbade anybody to own, read,

Baruch Spinoza

distribute, copy, or restate anything that he had written. Taking the hint, the Catholic Church also put his works on its Index of Forbidden Books.

When examining his writings, one understands why his proposals incensed the faithful in the seventeenth century. Spinoza

denied the immortality of the soul; strongly rejected the notion
of a transcendent God; and claimed that the Law (i.e., the commandments of the Torah and rabbinic legal principles) was neither given by God nor was any longer binding on Jews.[1] Thus it
is not surprising that Spinoza's theses were considered by Jewish,
Catholic, and Protestant contemporaries as monstrous sacrileges.

The strict ban on Spinoza by the Amsterdam rabbis was
quite in keeping with the customs of the time. To proclaim that
the Torah (the Old Testament) was not written by Moses and
that, therefore, Jews did not have to adhere to its precepts was
heavy medicine indeed. To deny the divine origin of the Jewish
Bible and its oral interpretations by the later rabbis had already
been deemed an abominable heresy in the Middle Ages by the
twelfth-century Jewish philosopher Maimonides, the most pro-
lific and influential Torah scholar of his time. Whoever denies
the written or the oral law, Maimonides had written, and if it be
only a single verse or a single word, was condemned forever and
ever for his great wickedness and sinfulness and would have no
share in the world to come.

"The world to come" is another source of contention because
a second tenet of Spinoza's worldview was that there actually is
no world to come: the soul dies along with the body. With this
Spinoza turned the belief system of all monotheistic religions
upside down. After all, the main theological reason for the belief
in the soul's immortality is that bad people who are not punished in this world, and good people who are not rewarded in
this world, get what is coming to them in the next. Hence, if the
soul dies with the body, an afterlife, a necessary requirement for
reward and punishment after death, is impossible.

As if these heresies were not enough, Spinoza doubled down
with one more outrage. In a most offensive sacrilege, he challenged the notion of God himself. True, God exists, Spinoza

maintained, but only philosophically. In his posthumously published work *The Ethics*, he went about proving God's philosophical existence—in a similar fashion to how Euclid proved geometric truths.[2] After stipulating eight definitions and seven axioms (self-evident statements that are obviously correct and do not need to be substantiated), he proves thirty-six propositions about God. Proposition XI, for example, states that "God . . . necessarily exists."

So, in concluding Part 1 of *The Ethics*, Spinoza summarizes: "I have explained the nature and properties of God. I have shown that he necessarily exists, that he is one: that he is, and acts solely by the necessity of his own nature; that he is the free cause of all things." But the god whose existence Spinoza proved and whose essence he described is not the God that Jews and Christians (and Moslems for that matter) worship. Rather, as Albert Einstein put it in a telegram to a New York rabbi, Spinoza's God "reveals Himself in the regular harmony of all that exists."[3]

To Spinoza, God and Nature are one; human emotions that rabbis and priests ascribe to Him play no role. To believe that good deeds are rewarded and sins punished is nothing but fantasy, Spinoza asserts, because everyday experience proves that good and bad luck befall the pious as well as the impious. The belief that rewards and punishments are postponed to the afterlife is nothing but naïve delusion. And to believe that storms, earthquakes, and diseases occur because God is supposedly angry is nothing but superstition. Unfortunately, however, true believers never give up on their entrenched prejudice; they prefer "to lump any contradictions together with other unknowns of whose use they were ignorant, and thus remain in their state of ignorance," rather than to abandon the whole fabric of their reasoning.

Not one to mince words, Spinoza did not let pass any opportunity to take a swipe at the rabbis who had excommunicated him: "Those persons whom the common people adore as the interpreters of nature and the Gods . . . know that amazement [about miracles] is their only means of arguing; hence, when ignorance is removed, their authority vanishes."

No wonder that the rabbis preferred their flock to remain ignorant.

43

APPLY FOR A PATENT OR
KEEP A SECRET?

Intellectual Property

After years of twiddling and fiddling, Dr. Daniella Düsentrieb has finally got her invention to work. (It could be a machine, a pharmaceutical compound, an electronic device, a piece of software, etc.) She is ready to commercialize her brainchild and will soon be able to reap the financial rewards. But first she must decide: keep the invention secret or apply for a patent? There are advantages and disadvantages for both paths.

Patents are meant to encourage innovation and enterprise. If inventors were not given the opportunity to profit from their inventions, nobody would invest time and effort in risky research that in many cases will lead to dead ends. Patents for successful inventions protect inventors from copycats for up to twenty years.

For national authorities who administer intellectual property to grant a patent, the invention must be useful, novel, and nonobvious. The latter condition means that the invention can't simply be an extension of something that is already publicly known or has already been patented. Examples of patentable inventions are machines, tools, computer hardware, software, games, processes to do something more efficiently, and even newly invented strains of asexually reproduced plants.

Many discoveries are not patentable. Laws of nature and scientific principles cannot be patented. Albert Einstein, formerly a patent clerk in the Swiss city of Bern, could not have patented relativity theory even though he discovered it. Nor are mathematical formulas patentable, although a specific usage of a formula may be. Substances found in nature are not patentable either, unless they are mixtures or combinations of multiple components. (Smokeless gunpowder, comprised of the naturally occurring elements potassium nitrate, charcoal, and sulfur, was patented and became the basis for the formation of the corporation DuPont de Nemours, Inc.) Genetically engineered corn may also be patented. Abstract ideas, on the other hand, may not.

If Dr. Düsentrieb's invention falls under the patentable items, she can apply for a patent. If she decides to do so, her application must describe the invention in all details. Once granted, the patent is published, and all particulars become publicly available. After all, potential competitors must be made aware of what they are *not* allowed to copy. A patent protects the invention for about twenty years, during which time Dr. Düsentrieb has the exclusive right to make use of it; if she permits others to exploit her invention, they must pay her royalties.

Dr. Düsentrieb can choose a different path: she can keep her invention and all details secret. Instead of describing the inner workings, revealing how her invention is assembled, and publishing all blueprints, she can just keep everything under wraps. Although less secure than patents, trade secrets also offer some measure of legal protection. They are considered intellectual property like any other possessions and their theft is a criminal offense.

But for an invention unprotected by a patent there exists a danger even greater than the theft of a trade secret. If competitors reinvent the trade secret on their own, or figure it out by lawful means—for example, by taking the device apart, studying it, and then producing it themselves—there are no grounds for

The Commissioner of Patents and Trademarks

Has received an application for a patent for a new and useful invention. The title and description of the invention are enclosed. The requirements of law have been complied with, and it has been determined that a patent on the invention shall be granted under the law.

Therefore, this   5,860,492

United States Patent

Grants to the person(s) having title to this patent the right to exclude others from making, using, offering for sale, or selling the invention throughout the United States of America or importing the invention into the United States of America for the term set forth below, subject to the payment of maintenance fees as provided by law.

If this application was filed prior to June 8, 1995, the term of this patent is the longer of seventeen years from the date of grant of this patent or twenty years from the earliest effective U.S. filing date of the application, subject to any statutory extension.

If this application was filed on or after June 8, 1995, the term of this patent is twenty years from the U.S. filing date, subject to any statutory extension. If the application contains a specific reference to an earlier filed application or applications under 35 U.S.C. 120, 121 or 365(c), the term of the patent is twenty years from the date on which the earliest application was filed, subject to any statutory extension.

Acting Commissioner of Patents and Trademarks

Attest

The United States of America

US Patent cover

a lawsuit; it is not illegal for competitors to reverse engineer a secret device. Worse still, the competitor may, in turn, decide to patent the original discovery himself, and then even the original inventor is no longer allowed to use it. In addition, in the course of litigation, more secrets may have to be disclosed.

So, the question facing Dr. Düsentrieb, and many corporations in similar situations, is whether to patent an invention or to safeguard it as a trade secret. If she foregoes patenting, she can use her invention immediately, without waiting for a long approval process . . . possibly only to find out that the invention is not patentable due to patent law restrictions. Especially in industries where product life cycles are very short, and the pace of innovation is fast, the time and cost of obtaining a patent might not be worth it: by the time a patent is granted, if at all, the product might already be obsolete. There is reason to forego patenting also if the life cycle is very long: a trade secret allows her to make exclusive use of her invention for much longer than twenty years. This is especially true for small companies where there is less reason to fear leaks.

However, if the product is easy to reverse engineer, or if employee turnover is great (thus increasing the risk of leaks), a patent is a better protection. An additional advantage of the patent route, especially for inventors who need funding, is that a prototype of an invention protected by a patent can be demonstrated to investors without fearing theft of one's ideas.

Lawsuits are nearly always costly and time-consuming, but patent infringements are easier to prosecute than theft of trade secrets. In both cases it is up to the inventor, however, to monitor for misappropriations and to take legal action. Although the situations are not always clearcut, patent owners only need to show infringement of their patent, whereas an inventor who did not patent the invention must demonstrate that the information was indeed a trade secret, that it is valuable, that it was kept secret, and that it was stolen.

44

DON'T THROW GOOD MONEY AFTER BAD

The Sunk Cost Fallacy

You have paid $200 to buy a ticket for a Bruce Springsteen concert that will take place Friday evening. On Friday you're tired; all you want to do is watch TV and go to bed. Moreover, the taxi rides would come to about $50. But no, you spent $200 to buy the ticket, so you'd better go. Otherwise the money is totally wasted.

Monday morning, at the board meeting, an investment proposal is discussed. Initial plans for construction of an office tower have been prepared by an architectural firm; the fee, $100,000, was paid last month. The financial officer lays out the situation: building the tower will cost ten million dollars. But, he warns, the rental market is down and the tower is a losing proposition. He advises that they abandon the project. "Oh no," the intern chirps in, "if we end the project now, the $100,000 will have been totally wasted."

ʒ❦

In both of these cases, the decision is obvious: don't fret about the past. Decide here and now whether it is advantageous to pursue the activity or project. Whatever money has been spent is gone; it is irrelevant, and you must ignore it. If a friend called you up and offered you a free ticket for tonight's concert, would you go or would you prefer to stay home and watch TV? If an architect offered your corporation ready-made plans for free, would it be profitable to invest in the project?

So, at the board meeting the financial officer tells the intern to shut up, and tells the other members that he doesn't even want to know how much has already been laid out because he doesn't care. The board members must wipe the slate clean and check whether it is profitable to go forward with the project as if they started from zero.

≈

The tendency for people and corporations to include past outlays in decisions about the future is called the sunk cost fallacy. Sunk costs are expenses that have already been incurred. They include all payments that have been made: salaries, insurance premiums, rentals, agency commissions, deposits, repairs, etc. When making decisions about the future, these costs must be ignored because they remain on the books regardless of whether one goes forward with the project or abandons it.

However, some down payments or contingency expenses may not be considered sunk if they are refundable. And if purchased objects or produced items can be sold, even if only for a fraction of their costs, these inflows *must* be taken into account for future decisions. Occasionally, outflows following the abandonment of a project, such as costs for the clean-up or charges for disposal, may surpass the loss that the project is expected to generate if it is continued. In such cases, it may be worthwhile to continue the project, not because of the sunk costs but because future costs are too high. Oh, and do not let an insurance contract lapse simply because you paid so many yearly premiums in the past (sunk costs) without ever getting anything in return.

છ•

The sunk cost fallacy is an example of a cognitive bias, a central topic in the field of cognitive psychology that studies how mental processes explain human behavior. Faulty methods of reasoning—using only information that is easily available, adjusting an estimate by comparing it to an anchor, answering a question depending on how it is framed, using rules of thumb instead of computations, or taking into consideration sunk costs—are considered cognitive biases, and they were studied in the early 1970s by two Israeli psychologists: Daniel Kahneman and Amos Tversky. After interviewing and questioning hundreds if not thousands of people about how they make decisions under uncertainty, they formulated a new branch of behavioral economics: Prospect Theory.

The two psychologists' work was instrumental in applying psychology to economic models of rational and irrational behavior. Kahneman was awarded the Nobel Prize in economics in 2002 for his investigations of how people make economic decisions. He and Tversky had done so by drawing on "cognitive psychology in relation to the mental process used in forming judgments and making choices." (Sadly, Tversky had died six years earlier.)

Why do business decision-makers occasionally, and private decision-makers often, fall into the trap of including sunk costs in their decisions? Behavioral psychologists have identified several reasons that may lead to this bias. One is that people are loss averse and may hold on to a bad investment because they do not want to feel bad about a loss . . . even though the realization will eventually catch up with them. Unrealistic optimism may be another reason; managers may believe (hope) that their investment plans will eventually pan out in spite of all contrary evidence. Yet another reason to fall into the trap is the burden of personal responsibility. A CEO is responsible to the shareholders for the decisions made. By writing off sunk costs as a loss and moving on, the CEO would implicitly admit to having made a bad decision.

On a private level, the trap looms everywhere: the reluctant concertgoer who ends up going anyway, reading a boring book to the end because one has already read half of it, finishing the dripping ice cream cone instead of tossing the cone into a bin, or staying to the end of a bad show instead of leaving during the intermission. Unwilling to admit that one has already wasted so much effort and time, one wastes even more.

Daniel Kahneman pointed to more weighty decisions: The sunk cost fallacy keeps people for too long in poor jobs, unhappy marriages, and unpromising research projects.

Enough said!

45

KABBALAH

Only Men Over Forty

K abbalah is a mystical and esoteric tradition in Judaism that tries to fathom the divine, the universe, the soul, and the relationship between them. Through metaphysical truths, it tries to grasp the essence of creation and cosmology and the nature of the infinite. Meditation, prayer, and the study of sacred texts enable believers to achieve a spiritual state that permits a connection to God.

At the center of the Kabbalistic conception is the Kabbalist's striving to reduce the gap between God and the physical world in which he is implanted, and to achieve emotional, experiential, and mental union with God. The entire physical reality and the phenomena contained in it are considered a "symbolic system" through which God reveals himself to the world. The Kabbalist tries to join this system through observation and to reveal the divine essence that is hidden everywhere and in everything that exists. In this way the Kabbalist tries to reach the heights of spiritual perfection, to return his soul to its divine root, and to cling to the Deity.

The word Kabbalah derives from the Hebrew word *lekabel*, which translates as "to receive." Rooted in traditional Jewish texts, such as the Torah (five books of Moses) and the Talmud (a compendium of oral laws and commentaries), the *Zohar*, probably written in the late thirteenth century by Moses de Leon in Spain, is considered the foundational text of Kabbalah.

According to Kabbalistic thought, the finite world was created from the "ein sof," the infinite, through a process of contractions. A central theme is the Tree of Life, consisting of the ten "Sephirot," the attributes through which God interacts with the world, wisdom, understanding, kindness, strength of judgment, beauty, eternity, splendor, foundation, and kingdom. Humans have a role in repairing the world spiritually and morally through actions, prayers, meditations, and the study of sacred texts.

The renowned modern philosopher Gershom Scholem lamented in 1941 that rationalist Jewish thinkers of the nineteenth century had little sympathy for Kabbalah. A result of this antagonism was that "all manners of charlatans and dreamers came and treated it as their own property. From the brilliant misunderstandings and misrepresentations . . . to the highly coloured humbug . . . the most eccentric and fantastic statements have been produced purporting to be legitimate interpretations of Kabbalism." As a result, Kabbalah "has become again what it was in the beginning: the esoteric wisdom of small groups of men out of touch with life and without any influence on it."[1]

Scholem would have been horrified at the Hollywood version of Kabbalah that has developed in the late twentieth century. Celebrities including Madonna, Demi Moore, Ashton Kutcher, Britney Spears, Lindsay Lohan, and Paris Hilton have become adherents of Kabbalistic thoughts and practices. The *Kabbalah Centre International*, founded in 1969 by Rabbi Michael Berg and his wife Karen, teaches courses in more than five dozen

אין סוף

Kether
כתר

Binah
בינה

Chochmah
חכמה

Da'ath
דעת

Gewurah
גבורה

Chesed
חסד

Tiphereth
תפארת

Hod
הוד

Nezach
נצח

Jesod
יסוד

Malchuth
מלכות

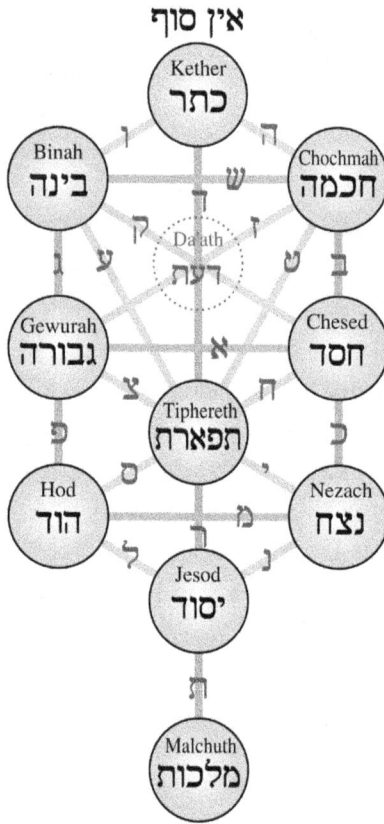

The Tree of Life (the ten Sephirot)

physical locations all over the world, as well as online, provides weekly astrology forecasts, and sells jewelry. According to the center's website, "despite continuous obstacles, Rav and Karen Berg successfully defied thousands of years of religious restrictions, opening the doors for everyone to study Kabbalah."

One of the obstacles to which the website refers is that tradition limited teaching Kabbalah to men over the age of forty who are well-versed in Torah, Talmud, and Jewish law. The Bergs' decision to make Kabbalistic teachings accessible to people of all ages, genders, and backgrounds directly defied this long-standing norm.

Traditionally, the reason men younger than forty were supposed to be kept ignorant of Kabbalah teachings was because only when a man (yes, no women) reaches age forty was he considered sufficiently mature to deal with Kabbalah's mystical nature, its complexity, and its societal structures. Younger men who are prone to misunderstanding the deep metaphysical concepts could be led to confusion, anxiety, despair, and frustration.

Because it takes years of study of the Torah, Talmud, and Jewish law to immunize men from the dangers of heretical interpretations, a shallow understanding of the divine mysteries, cosmology, and spirituality could have profound psychological consequences. Individuals might develop irrational fears of divine retribution or misinterpret symbolic teachings as literal realities.

Other kinds of problems may arise. Immersing oneself in esoteric teachings without proper guidance can isolate a young person from his community. Immature men may experience cognitive dissonance and detach themselves from reality. Such detachment can lead to behaviors that alienate them from family, friends, and their religious peers. Belief in spiritual powers might escalate to a messianic complex.

Moreover, existential questions about purpose, free will, and the meaning of life can destabilize someone who is not yet ready for profound introspection. It may lead to personal doom or cosmic hopelessness. Obsessive behavior, focusing on spiritual practices and seeking mystical experiences, can dominate one's thoughts and actions. Mystical practices may exacerbate mental health problems such as anxiety and depression.

What led Kabbalists of olden times to keep men under forty ignorant of the teachings of Jewish mysticism was that these teachings were considered too potent, complex, and transformative for an unprepared mind. Meditative practices without proper grounding might induce dissociation, irrational fears, paranoia, or distorted views of the world.

Restricting the study of Kabbalah to older and experienced individuals was meant to minimize the risks of misuse.

IV

REFUSE TO KNOW

I t would seem rational that one should prefer more knowledge rather than less. A most confusing aspect of epistemology is that this is not always so; occasionally knowledge is ignored willingly. Some truths may challenge deeply held beliefs, others may be hurtful or distressing, and some may be too difficult to comprehend. Better not to know . . . and remain in blissful ignorance.

46

MURDER, SUICIDE, OR
THE GODS' WRATH?

The Square Root of 2

I s it conceivable that someone would be punished by death because he did not want to keep humankind ignorant of certain numbers? To keep mathematics from advancing? According to numerous legends—*nota bene*: they are legends—this is what is supposed to have happened in the fifth century BCE. Details, vague as they are, were only described eight centuries later by the Arab philosopher Iamblichus and may be somewhat, or completely, untrustworthy. But the story is too good to pass up, so I shall recount it with all the necessary caveats.

In around 530 BCE, the Greek philosopher Pythagoras founded a sect. Gathering a group of men around him, he engaged them in philosophical discussions and lots of humbug. As one modern writer put it, "some were primarily interested in scientific and intellectual questions while others wallowed in the taboos, obscure sayings and superstitious guides to life that made up the religious side of the Pythagorean curriculum."[1] We shall focus on the former group who occupied themselves with observing and studying astronomy, mathematics, and music.

Playing around with a monochord, a musical instrument that has a single string tautly attached at both ends, the Pythagoreans made a momentous discovery. The hums that the string emits when made to vibrate are especially pleasant-sounding when one sound was followed by another with the string held down halfway, or two-thirds of the way, or three-quarters of the way. In a leap of faith, they concluded from this discovery that all of nature is governed by numbers. By number they meant integers such as 1, 2, 3, and fractions such as ½, ⅔, ¾, which are collectively known as the rational numbers. "All is number" became the Pythagoreans' motto. So far, so good.

At the same time, the members of the sect were concerned with another important finding, the so-called Theorem of Pythagoras. Generally attributed to the master himself, it may have been one of his disciples who formulated it. Anyway, the eponymous theorem, taught today to schoolchildren all over the world, was already known to the Chinese and the Babylonians at least one thousand years before Pythagoras was even born. For example, in 1894 a French archeological expedition excavated a Clay tablet dating to between 1900 and 1600 BCE in what is now Iraq. This tablet was apparently used by ancient land surveyors to draw accurate boundaries. The cuneiform markings instructed the surveyor on how to make accurate right-angled triangles.

What the Babylonians and the Chinese discovered, and the disciples of Pythagoras rediscovered, was that in a right triangle the square of the long side (the hypothenuse) is equal to the sum of the squares of the short sides (the legs). If the legs are denoted by A and B, and the hypothenuse by C, then $A^2+B^2=C^2$. The Pythagoreans verified this with various triangles, and it always worked: if the legs are 3 and 4, the hypothenuse is 5 ($3^2+4^2=5^2$); if the legs are 8 and 15, the hypothenuse is 17 ($8^2+15^2=17^2$); if the legs are 5 and 12, the hypothenuse is 13.

But then—according to the narratives and legends—something happened that blew the sect's entire belief system asunder. One of Pythagoras's disciples, a man by the name of Hippasus of Metapontum, wondered how long the hypothenuse of a triangle

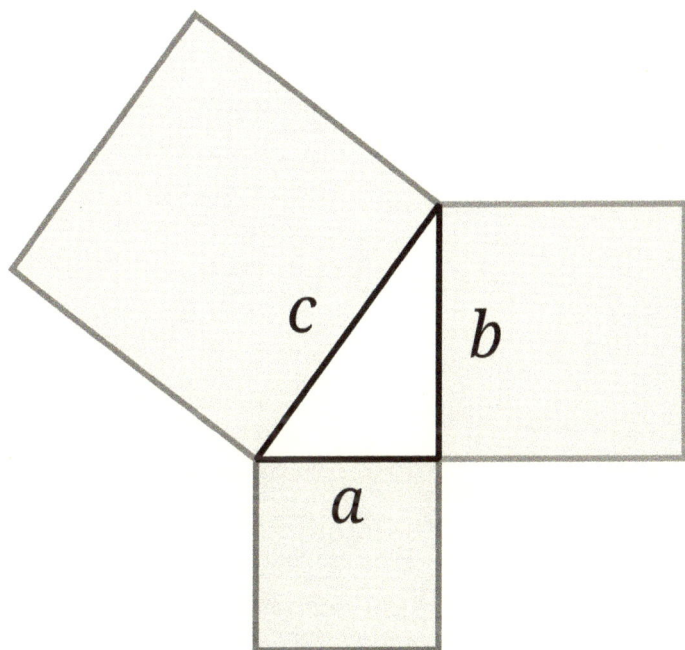

The Pythagorean Theorem

is if both legs are one unit long. He figured that if A and B are equal to one, then C must be some number whose square is 2. But try as he might, Hippasus could find no fraction that would satisfy $1^2 + 1^2 = C^2$. Although $1\frac{9}{7}$ comes fairly close, it is still far off the true value of C, namely, 1.414213 . . . In fact, Hippasus or one of his colleagues soon found out that no fractional number exists whose square is 2. The proof is fairly simple (see the appendix to this chapter).

The discovery of a so-called irrational number—a number that is neither an integer nor a fraction—proved that things existed in the universe that could not be comprehended through rational numbers. To the Pythagoreans this was earth-shattering.

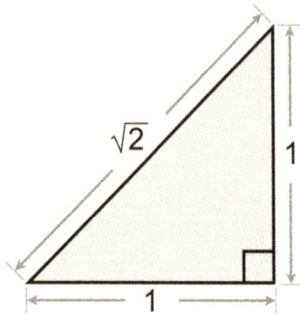

Square root

If only integers and fractions were considered as numbers, then the existence of irrationals proved that "all is number" was no longer valid. Translated into modern terms, this was definitely *TS/SCI* stuff, *top secret/sensitive compartmented information* (see chapter 4). Leakage of this information would do irreparable damage to the Pythagorean worldview.

To keep up the appearance that their Weltanschauung was intact, the fact that irrational numbers exist would have to be kept under wraps; the world must never know about them. But how could the information be stopped from getting out? Hippasus, proud or upset about his finding, would certainly blabber about it. And here the narratives, unreliable as they are, diverge even more. Some say that Pythagorean enforcers simply killed Hippasus. Other sources maintain that the distraught man, whose entire belief system had been shattered, committed suicide. Still others claimed that the gods took revenge and caused Hippasus to fall off a ship and drown.

ॐ

So much for the legends; ancient sources for what actually happened are very thin. Walter Burkert, a scholar of classical languages at the University of Zürich and an authority on the Pythagoreans, was convinced that it was "the tradition of secrecy, betrayal, and divine punishment [that] provided the occasion for the reconstruction of a veritable melodrama in intellectual history."[2] In fact, he claimed, what gave rise to at least some of the legends may have been a matter of mistranslation: the Greek word for "irrational" (*arretos*) also means "secret" or "unsayable." Hence "the 'ineffable because irrational' is at the same time the 'unspeakable because secret', " he wrote.

In any case, the secret, should it have been such, could not be kept. Indian mathematicians had been aware of irrational numbers as early as the eighth century BCE, and they and Muslim mathematicians used them throughout the ages. A rigorous formalization of the real numbers (the ensemble that encompasses both rationals and irrationals) had to wait, however, until the nineteenth century, when several German mathematicians put arithmetic on a solid footing.

In 1891, a German mathematician, Georg Cantor, discovered something strange. Although there are infinitely many rational numbers and infinitely many irrational numbers, there are more irrationals than rationals. That's strange: no wonder the Pythagoreans were confused. Unfortunately, Cantor's contemporaries were also very confused and—partly due to their hostility toward his theories—this mathematician spent the remainder of his life in a psychiatric sanatorium.

Are there different degrees of infinity? The answer is yes. As numerous as the rationals are, they represent just a blip in the ocean of all real numbers. In fact, if you throw a dart at the numbers line, you are bound to hit an irrational number. The probability of your dart landing on a rational number is *zero*, the probability of hitting an irrational number is *one*. (Somewhat counterintuitively, this does not mean it is impossible to hit a rational number; it just means that the probability is zero.)

But throwing darts is no way to prove a mathematical fact, and Cantor showed in a rigorous fashion that the rational numbers can be counted whereas the irrationals cannot be counted: when depicted in a grid, a counter moving vertically, diagonally, and horizontally will eventually reach every rational number. Hence rational numbers are "countably infinite" and the irrationals are uncountably infinite.

According to the legends, Hippasus was on the verge of revealing the beauty of irrationals to the world. The enforcers . . . or the gods . . . preferred to keep the world in ignorance.

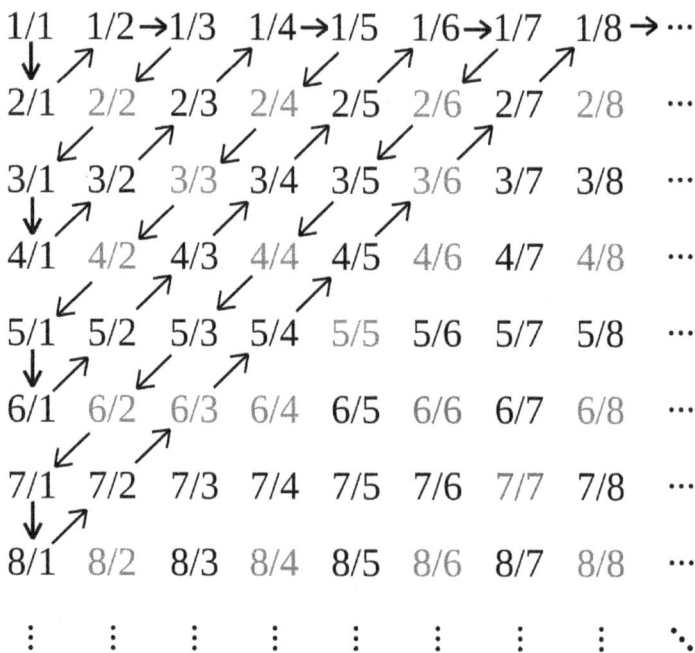

Georg Cantor's method of counting rational numbers

APPENDIX

If the square root of two were a fraction, we would have,

$$\sqrt{2} = p/q,$$

where either p or q must be an odd number. (If both were even, the fraction could be reduced to ${}^{p/2}/_{q/2}$.) Hence

$$2 = p^2/q^2 \qquad \text{or} \qquad 2q^2 = p^2.$$

Hence p^2 is an even number, which means that p must be even. (Only the squares of even numbers are even.) But if p is even, p^2 must be divisible by 4, which means that q is also even. But since, as I said, p and q cannot both be even, we have a contradiction. Hence, no fraction exists whose square is equal to 2. QED

47

"EPPUR SI MUOVE!"

The Trial of Galileo

That planet Earth is a sphere was already known to the Ancient Greeks in the fourth century BCE. Appropriate evidence cited by Aristotle were the facts that ships disappear hull first when they sail over the horizon and that Earth casts a round shadow on the moon during a lunar eclipse. That is as far as correct observations went at the time. (See, however, chapter 51.)

Given people's belief in their supremacy, it is no wonder they believed that the planet on which they lived, planet Earth, must be the center of the universe. The most prominent expression of this worldview was given in the *Almagest*, a book written by the Greek astronomer Ptolemy in the second century CE. The geocentric model that Ptolemy proposed, which prevailed for twelve centuries, placed the Earth in the center of the cosmos, as the word "geocentric" indicates. All other planets and the Sun rotated around it.

The church was only too happy to adopt this worldview. At the center of everything is God; God created man; man lives on Earth; hence Earth must be the immobile center of the universe. The fact that, to earthly observers, Mars appears to move forward, then stop and move backward, then stop and change direction again and move forward, was just one of God's mysterious ways.

Persian and Islamic scholars had their doubts about a geocentric universe, but it was the Polish astronomer Copernicus (1473–1543) who finally put the Sun squarely at the center of the universe (pun intended). With the Sun motionless in the center, the heliocentric model accorded much better with the planets' orbits. Although the model clashed with Catholic doctrine, church fathers did not take much notice because Copernicus's work, *De revolutionibus orbium coelestium* (On the revolutions of heavenly spheres, first published 1543), was written in Latin and hence incomprehensible to the general population.

It was more than half a century after that, that before the Italian astronomer and mathematician Galileo Galilei (1564–1642) and his German peer Johannes Kepler (1571–1630) took up the cudgel. Galileo was not just an extraordinary scientist but also a gifted tinkerer. He had designed and built the most powerful telescope of his time and used it extensively for astronomical observations. This supported his conviction that the Sun was indeed in the center of the universe. Kepler, more a theoretician than an observer, discovered the three laws of planetary motion, which explained the observed elliptical shape of planetary orbits. (That the orbits were not circular was a fact Galileo had preferred to ignore.)

In 1632, Galileo published *Dialogo sopra i due massimi sistemi del mondo* (The dialogue concerning the two chief world systems), in which he makes the case for the heliocentric system. Written in Italian, not in obscure Latin, the *Dialogo* became quite popular among the general readership. This time, the church did take notice. Pope Urban VIII's annoyance was especially vehement.

According to contemporary accounts, the pope used to be a warm, compassionate, intelligent human being. As a cardinal, he was well inclined toward Galileo, and even after his election to the papacy, Urban VIII granted Galileo six audiences, each lasting more than an hour. He was okay with Galileo's writing and lecturing on the heliocentric model as long as he treated it not as fact but as a hypothetical idea.

Lulled into a false sense of security, Galileo began working on the *Dialogo*. To satisfy the church, the book had to present arguments also in favor of the Ptolemaic system. Set up as a discussion between the scientist Salviati, the wise Sagredo, and the simpleton Simplicio, Galileo complied with the requirement, but he shrewdly let Simplicio take the defense of the geocentric system. Although the implication should have been clear, it was, at first, lost on the church censors. When Galileo traveled to Rome in 1630 to obtain permission from the papal inquisitor to publish his work, the subterfuge escaped the censor's scrutiny, and Galileo had the book printed in Florence.

But the respite was only temporary. In July 1632, the inquisitor in Rome instructed his colleague in Florence to prevent distribution of the work; in September the pope summoned Galileo to Rome. Accused of insubordination, an investigation began the following April. After the first hearing, Galileo was put under arrest. On April 22, at the second hearing, Galileo admitted to having erred. On May 10, he submitted his defense and a plea for grace, given his failing health.

The heresy trial took place on June 22, 1633, in a Dominican monastery. Referring to the dialogue form of his book, Galileo denied having taught the Copernican system. But threats of torture and even burning at the stake convinced Galileo to renounce his lifelong teaching that the Sun was the center of the universe and that Earth moved around the Sun. Kneeling in front of the tribunal, Galileo assured the cardinals that he would "altogether abandon the false opinion that the Sun is the center of the world and immovable, and that the Earth is not the center of the world, and moves," and committed himself "with sincere heart and unfeigned faith" to "abjure, curse, and detest the aforesaid errors and heresies, . . . and I swear that in the future I will never again say or assert, verbally or in writing, anything that might furnish occasion for a similar suspicion regarding me."[1]

It did not help; Galileo was declared guilty for his public stand as a Catholic who tried to convert the Church to the new science. He was spared burning at the stake, but he was condemned to house arrest for the rest of his life. In addition, he was to recite the seven penitential psalms once a week for three years. The latter task was taken on in his stead by his daughter, a nun.

But Galileo's heart was not as sincere, and his faith not as unfeigned, as he had been forced to pretend under threat of torture and execution. Upon exiting the trial chamber, he is said to have muttered under his breath *"eppur si muove!"* (and she, the earth, moves nevertheless).

Pope Urban VIII and his acolytes had done their very best to keep mankind ignorant of the true properties of the planetary system. But not everybody agreed with the verdict; of the ten cardinals who sat in judgment over Galileo, three did not sign the judgment. And in 1992, more than three-and-a-half centuries later, at a ceremony in Rome, the Pontifical Academy of Sciences closed a thirteen-year investigation and overturned the verdict with a formal rehabilitation of the great astronomer. Pope John Paul II officially declared that Galileo had been right.

48

DIAMONDS ARE A GIRL'S BEST FRIEND

A Mineral Consisting of Pure Carbon

I n the 1953 film *Gentlemen Prefer Blondes*, Marilyn Monroe, playing the gold-digging showgirl Lorelei Lee, sang the unforgettable song *Diamonds Are a Girl's Best Friend*. What she expressed was that love is a luxury whereas diamonds provide a more tangible and lasting value to a girl.

But why, of all things, diamonds?

Created three billion years ago, diamonds are pure carbon that has been compacted at enormous pressures and temperatures at depths of between 150 and 250 kilometers under the Earth's surface. Volcanic eruptions eventually carried them to the surface.

Diamonds have remarkable optical, thermal, chemical, and electronic properties: they are excellent electrical insulators; have the highest thermal conductivity of any material, yet barely expand when heated; are transparent to ultraviolet, visible, and infrared light; and are chemically inert to nearly all acids and bases. Diamonds are the hardest and stiffest material known; hence they are very useful as cutting and polishing tools. Most important, they are also beautiful: when they are themselves cut and polished, they display extraordinary luster, sparkle, and brilliance.

But other than that, they have no intrinsic value. So the question is: What is their tangible and lasting value to girls like Lorelei Lee?

Like the price of nearly everything, the cost of diamonds is determined by demand and supply. Compared to other gem stones, diamonds are plentiful. The De Beers Group, the South African mining firm that controls the majority of the world's diamond supply, maintains vast stockpiles and tightly controls supply. But the process of diamond extraction from the mines is labor-intensive, and it is difficult to find a large piece of flawless diamond. And then there's social pressure on young men that forces them to buy diamond engagement rings for their fiancées that cost as much as what they make in several months' salaries. De Beers's famous advertising slogan "A Diamond Is Forever," coined in 1947, never lost any of its power.

But in recent decades all that has changed. Diamonds are just pure carbon that has crystallized under intense pressure and heat, so it should be possible to produce them artificially. Thus, unlike medieval alchemists who tried to produce gold out of cheap metals, modern scientists based their efforts to produce diamonds on a solid theoretical basis.

Attempts to create diamonds artificially began in the late nineteenth century. At first, the chemists, pharmacists, and physicists either did not succeed at all or produced something that then turned out not to be a diamond. This changed in the 1950s when the American General Electric Company got serious. Building on research by Percy Bridgman of Harvard University, who had received the Nobel Prize in physics in 1946 for his research on the properties of matter under high pressure, GE developed machinery that mimicked the high pressure, high temperature (HPHT) conditions of natural diamond formation in the Earth. Under pressures of between sixty and one hundred tons per square centimeter—that's the weight of about eighty passenger cars concentrated on a dime—and temperatures above two thousand degrees centigrade, the scientists created a minuscule synthetic diamond from carbon material. Although too small and too imperfect for jewelry, it did herald synthetic diamonds for industrial usage.

Production by several start-up companies of larger, gem-quality crystals began in the mid-1990s and became increasingly successful. (Another process to grow synthetic diamonds is by chemical vapor deposition [CVP]: a small seed of a diamond crystal is taken and grown from a hydrocarbon gas mixture layer by layer in a chamber.) In 2018, the DeBeers Group decided that if they can't beat them they'd join them: they introduced a new brand of jewelry called "Lightbox." Their synthetically created diamonds sell for about $2,000 for a one carat stone, a fraction of the cost of the naturally occurring counterpart. Another advantage that diamond traders generally do not talk about—to avoid undermining their original market—is that lab-grown diamonds are ethically produced because no slave laborers are abused to mine them. And the best part: practically nobody can tell the difference!

The general buyer certainly can't, and neither can most jewelers. Synthetically grown diamonds are chemically, physically, and optically identical to diamonds produced by geological processes. Very sensitive infrared spectroscopy and X-ray absorption spectroscopy technologies are required *maybe* to detect trace amounts of defects in the atomic lattice structure of diamonds. The Gemological Institute of America has developed highly sophisticated techniques and instruments that *may* distinguish synthetically produced diamonds from those that were created geologically.[1] De Beers labels their Lighthouse stones with a tiny laser-engraved logo so as not to cannibalize its core business of mined stones.

Keeping up with the times, the Federal Trade Commission (FTC) modified its definition of a diamond in 2018. In 1956 a diamond was described as "a natural mineral consisting essentially of pure carbon crystallized in the isometric system," but after much discussion the FTC omitted the word "natural" from the definition in 2018. To paraphrase Gertrude Stein, "a diamond is a diamond is a diamond."

And now the question is: If a lab-made diamond is indistinguishable from a mined diamond, does it matter whether the stone came about by a geological or by a technological process? May one remain ignorant about the circumstances of the stone's emergence? After all, it matters little whether an orchid was grown in a hothouse or in southern Patagonia. And ice is real whether the water froze in your freezer or in the Swiss Alps.

The question can be answered by recalling the famous ring parable in the eighteenth-century play *Nathan the Wise* by the German writer Gotthold Ephraim Lessing. For many generations a magical ring had been passed from father to favorite son. After several generations, a father had three sons whom he loved equally. So he had two replicas made and bequeathed one ring to each son. After the father's death, the three men quarreled about who owned the true ring until a wise judge advised them to believe and behave as if each one of them owned the true ring.

The mystique of geologically sourced diamonds is anything but rational. For some people, spurred on by inane slogans such as "an Earth mined diamond is nature's form of art" or non sequiturs such as "the one-of-a-kind gemstone still remains a woman's best friend," it is precisely the high cost of mined diamonds that attracts them. This is what became known as the Veblen Effect, named after the Norwegian-American economist Thorstein Veblen (1857–1929) who described it in his 1899 work *The Theory of the Leisure Class*.[2] So-called Veblen Goods defy the standard law of demand, which states that as prices increase demand for regular goods decreases. For Veblen Goods, demand increases as price rises because the high price itself makes the product more desirable as a symbol of wealth, status, or exclusivity.

But for many, it is not the price but the diamond's beauty and sparkle that is the allure. And in that case, one may as well remain ignorant about the stone's origin.

49

THE SILENT EPIDEMIC OF MISUNDERSTANDINGS

Pluralistic Ignorance

Scenario 1: Alexandra participates at her first condo board meeting. Topmost on the agenda is a decision about whether dogs should be allowed to run around in the lobby. Alexandra does not like dogs at all; neither does Bruce, her neighbor from across the corridor; nor do Carla, the lady from the twelfth floor, and David from the fifth floor. The majority of the members of the board dislike dogs with the notable exception of Emma, the chairwoman who owns two Dobermans, and Fred, the proud owner of a Pitbull. Emma opens the meeting by professing her love for dogs in general. Nobody else speaks up because love for doggies is generally considered the norm. When the time comes to vote, the result is a unanimous vote for allowing dogs into the lobby.

Scenario 2: Many a school kid starts smoking, coughing and gagging, because his friends profess how great it is. Frat brats get drunk every weekend, because the bros seem to enjoy being sozzled. Deep down the smokers and the drinkers would prefer to abstain, but because the majority around them inhale and imbibe they follow suit.

Scenario 3: Come election time, a majority of voters cast their ballot for the candidate whose ubiquitous posters and incessant

TV spots seemed to indicate that everybody wants her. In the mistaken belief that the majority support her, many who don't, vote for her anyway.

≥▲

To describe situations where many persons in the group may privately reject a belief or behavior but mistakenly assume that most others in the group accept it, the term "pluralistic ignorance" has been coined. In essence, when sufficiently many individuals in a group are misinformed about the true sentiments, be it due to social dynamics, group pressure, or ignorance of the true prevailing sentiment, collective misunderstandings arise.

Usually such misunderstandings arise when individuals hesitate to express their true thoughts and feelings out of fear of social disapproval. To gauge what the norm is, they rely on public cues and observed behavior by others, while at the same time hiding their own opinions. But the cues may be ambiguous, and dissenters are mistakenly led to conform to what they believe is the majority opinion. Thus the false illusion that the majority holds a particular belief is perpetuated.

True, many groups, especially those that work together or live in the same neighborhood, are more or less homogenous, with most members having similar likes and dislikes. But sometimes a vocal minority convinces the silent majority to go along with their opinion because the less vocal majority falsely believes that the loudly proclaimed views are the opinion of the majority.

≥▲

Misperceptions due to pluralistic ignorance can significantly affect group dynamics, decision-making, and societal norms.

The vocal majority

Intimidated by the pressure to conform, individuals fear to stand out or appear deviant. This leads to a cycle of conformity and self-censorship, thereby reinforcing the illusion of a consensus. It can perpetuate harmful or undesirable behaviors that are mistakenly perceived to be the accepted standard; individuals may engage in them to conform, even if they privately disapprove.

The phenomenon has wide-ranging consequences for personal relationships, group behavior, and public policy. It stifles dissent, which leads to a lack of diversity in perspectives and ideas and hinders innovation, problem-solving, and the development of critical thinking skills. Pluralistic ignorance can perpetuate harmful or undesirable behaviors that are mistakenly perceived to be the accepted standard; individuals may engage in them to conform, even if they privately disapprove. Policymakers may make decisions based on a perceived majority opinion that is not truly representative of public sentiment.

To counter the marginalization of nonconformist thinking that results from pluralistic ignorance, one must ensure that the voices and opinions of the perceived minorities are also heard and valued. As soon as emboldened individuals begin to question and challenge societal norms, the cycle of conformity that perpetuates pluralistic ignorance is broken. Hence, to break the progression of pluralistic ignorance, open communication must be encouraged and environments must be created in which individuals feel bold enough to share their true beliefs and opinions, for example, in workshops, town hall meetings, or online forums.

50

"NON, JE NE REGRETTE RIEN"

Anticipated Regret

I n one of her most famous songs, *Non, je ne regrette rien*, Edith Piaf, the celebrated French singer of the 1950s and 1960s, sends a message of optimism. Looking back at her life, the forty-four-year-old performer (she would die only three years after making the song a hit) affirmed that she was at peace with everything that has happened to her, both good and bad.

> *Non, je ne regrette rien,*
> *Ni le bien qu'on m'a fait,*
> *Ni le mal . . . tout ça m'est bien égal.*

(No, I regret nothing, neither the good that was done to me, nor the bad . . . it's all the same to me.)

In general, when deciding among several courses of action, one weighs the benefits of each (*le bien qu'on m'a fait*) against the costs (*le mal*). They could be financial success or loss, well-being or depression, happiness or sadness. With hindsight, most people are not as nonchalant as Edith Piaf pretended to be; they *do* regret certain decisions they made in the past. What about regret of decisions that you have not yet made?

Edith Piaf

Surprisingly, regret also plays a role when looking forward. Regret about a decision made today may occur, if at all, only at a later time, but most people factor it into their present choices. This is called "anticipated regret." In particular, anticipated regret shapes decisions about whether to obtain knowledge or remain ignorant about something. Faced with the decision either to acquire knowledge or to avoid it, people often take into account how they might feel afterward about their choice.

Received wisdom has it that information is always beneficial. After all, one would think quite rationally that knowledge cannot be detrimental, because if it were, one could simply discard it. Unfortunately, this is not true. Once it has registered in one's mind, information cannot be ignored. And some pieces of knowledge (for example, the date of one's death, the fact that a man is not the father of a child he thought was his, or the news that one's spouse is having an affair) impose emotional costs. So, whenever the distress that may be caused by some information is greater than the balance of *le bien* minus *le mal*, one had better remain ignorant. Moreover, in some cases ignorance may actually evoke emotional benefits, as you will see next.

&.

In one research project, it turned out that some 10 percent of Canadian adults with a family history of Huntington's disease chose not to have a predictive test for the deadly disease. Just like the Trojans ignored the prophesies of the mythological priestess Cassandra, who was granted the power of foretelling the future, one tends to avoid potentially bad news, particularly when one has no means of preventing the predicted, unfortunate outcome.

Another study showed that 20 percent of Malawi adults at risk for HIV chose not to learn about the results of an HIV test even when offered monetary incentives. More recently, why did many people who felt ill choose not to test themselves for COVID-19? Medical researchers often link not wanting to know with self-deception, dishonesty, and—especially in the COVID-19 example—shirking of responsibility.

The psychologists Gerd Gigerenzer, director of the Max Planck Institute for Human Development in Germany, and his Spanish colleague Rocio Garcia-Retamero sought an explanation for why people willingly ignore information, thereby foregoing knowledge that is available or that could easily be obtained. They questioned one thousand people each in Germany and Spain about their attitudes to knowing or ignoring. The results in both countries were similar. (I report here the results for the German sample.)

First, questions were asked in the context of negative events:

Would you want to know today when your partner will die?
No: 89.5 percent.

Would you want to know today from what cause your partner will die? No: 90.4 percent.

Would you want to know today when you will die? No: 87.7 percent.

Would you want to know today from what cause you will die?
No: 87.3 percent.

Would you want to know whether your marriage will end in divorce? No: 86.5 percent.

Between 85 percent and 90 percent of the German and Spanish test subjects preferred ignorance for negative events. "A majority of people anticipate that they would regret an unpleasant answer—that their partner will die very early, that they will die early, or that their marriage will end in divorce. This anticipated regret is larger than the regret associated with foregoing possibly good news. As a consequence, they prefer deliberate ignorance."[1]

Knowledge about the outcome of positive events was also spurned by the German and Spanish test subjects, but less so than for negative events: between 40 percent and 70 percent prefer to remain ignorant of positive events.

Would you want to know how a video-recorded soccer game ended, while you watched? No: 76.9 percent.

Would you want to know in advance what you are getting for Christmas? No: 59.6 percent.

Would you like to know whether there is life after death? No: 56.9 percent.

Would you have a sapphire tested to see whether it is genuine?
   No: 48.6 percent.
Would you want to know the gender of your child before birth?
   No: 40.3 percent.

Gigerenzer and Garcia-Retamero concluded that for positive events "people anticipate regret in knowing the answer because this would spoil surprise and suspense."

৯৯

Hence deliberate ignorance can be explained by anticipation. For negative events, deliberate ignorance avoids the regret that would arise if bad news should prove to be true. For positive events, deliberate ignorance avoids the regret that comes when premature knowledge spoils the punch line.

Edith Piaf notwithstanding, regret is by no means *bien égal*.

51

ILL-INFORMED BUT HARMLESS?

Flat-Earthers

An ant crawling on a bowling ball may be forgiven for believing that it is crawling along a table top. With its limited range of view, the surface of the round bowling ball just seems flat. A fly buzzing above the bowling ball, on the other hand, realizes that the object is a sphere. The ant's erroneous worldview derives from the fact that it only perceives the ball's two-dimensional surface, and only a small part of it. The minute curvature of the three-dimensional spherical bowling ball escapes it. The fly gets the full picture: buzzing in the air above, it perceives the three-dimensional sphere.

In contrast to ants, humans have known since ancient times that the Earth is a sphere, even though they too walk about on a surface that seems flat. The Greek geographer and astronomer Eratosthenes had even estimated the Earth's circumference in the third century BCE to within 3 percent of the true value. That was much closer than the estimation of Christopher Columbus who also realized that the Earth was round but severely underestimated its size. That the Earth is a sphere (more exactly: a geoid, i.e., a sphere, flattened at the poles) has been received wisdom for millennia.

It is all the more surprising, then, that even in the twenty-first century there are multitudes of people who, like ants, are convinced that the Earth is flat. Just look around you, they implore the round-earthers, and trust your senses. Do you see any curvature? Do you really believe that there are people on the other side of a globe, hanging onto Earth upside down? Just do your own research . . . mainly by watching YouTube videos of like-minded flat-earther colleagues.

There's even an association, the *Flat Earth Society*, aka the *International Flat Earth Research Society*, complete with membership cards, books, newsletters, and international meetings. They prefer to ignore that the Earth is a sphere; their worldview maintains that the Earth is a flat disk with the north pole in the

The Flat Earth Society logo

middle. This disk is surrounded on all sides by glacier mountains, called Antarctica, that hold back the oceans.

By the way, flat-earthers also believe that our planet is stationary, with the Sun circling around it at a height of five thousand kilometers (the true distance is one hundred fifty million kilometers), that the disk of the Earth is lifted by dark energy, that what is "commonly referred to as gravity" is actually the force of the flat Earth pushing everything upward, that the Coriolis Effect is fictitious, that Foucault's pendulum is given a lateral push at the start, and that there is an infinite plane of exotic matter somewhere below the disk. What is beyond the ice walls of the Antarctic is the subject of ongoing "research."

The fact that ships traveling away from the shore disappear behind the horizon—the classic proof of the Earth's curvature that convinces every high-schooler—is explained away by ocean swells, limited optical resolution, and the phenomenon of refraction, which supposedly causes objects normally on the horizon to sink below it. However, this very phenomenon of refraction is dismissed as an explanation for faulty observations carried out by flat-earthers along a six-mile length of a river. (A flag was still visible at that distance, although according to round-earthers it should have sunk below the horizon. The reason it did not was due to . . . yes, refraction.)

Now, one would think that even if the numerous proofs of the Earth's spherical shape, from Eratosthenes to Einstein, do not convince skeptics, the pictures from outer space, the accounts of astronauts, and films of the moon landing would certainly do it. But no! Pictures have been altered, reports are fake, the moon landing was simulated by actors. Whereas blurry pictures of an airplane flying *behind* the Sun are the real proof of a low-hanging Sun. And so on, and so on. Whoever claims to have personally discerned curvature, say from an airplane window, must have been experiencing a form of confirmation bias.

Apart from fundamentalist flat-earthers who derive their conviction from the Bible, the flat-earth concept is not really tied to any ideology, group identity, or economic interest; it just signals willful ignorance derived from vague, ill-informed, anti-intellectualism. Mainly it indicates a deep distrust of politicians, federal agencies, scientific institutions, and academic researchers.

It is interesting that most flat-earthers do not disdain science as such; it is only the scientists that they scorn. In fact, they attempt to "prove" their worldview, basing it on pseudoscientific theories (which are easily refuted), elaborate experiments (that they interpret incorrectly), and photographic evidence (that they dismiss if it contradicts their worldview). The underlying problem is one of mistrust in the so-called elites rather than mistrust in physics.

According to flat-earthers, everything to do with space travel is one great conspiracy. And who's in on it? The CIA, the FBI, NASA, and all international space research institutes. To what purpose? "The purpose of NASA is to fake the concept of space travel to further America's militaristic dominance of space." Flat-earthers note that NASA is not conspiring to portray the Earth as round; its scientists portray it as round because they truly think it is round. But because nobody ever was in space, nobody landed on the moon, and space travel never happened, the people at NASA cannot know what the shape of the Earth actually is.

৺

Flat-earthers may be considered innocent and harmless, although medievally ill-informed—with apologies to medieval people who *were* aware of the round Earth. Sadly, whenever and wherever conspiracy theories flourish, antisemitism is often not far behind. And so it is with some (not all) flat-earth ideologists. Even though no conceivable connection between the shape of the Earth, the Jews, Zionism, and the holocaust can be made, the alleged global conspiracy that surrounds the so-called round-earth-hoax presents an opportunity to spout antisemitic rhetoric. When some (again: not all) flat-earthers rail against globalists, they do not refer to people who believe the Earth is a globe; they use the term as a dog whistle to invoke antisemitic tropes.

52

DON'T EVEN TALK ABOUT IT

Critical Race Theory

I n the 1960s and 1970s, the critical legal studies (CLS) movement in the United States began to examine how laws are
intertwined with social issues. Proponents of CLS believe
that the law supports the interests of those who create it.
Marxist-oriented scholars and social activists, but not only they,
determined that the law benefits the wealthy and powerful at
the expense of the poor and the marginalized.

From these roots an academic field emerged two decades
later, called critical race theory (CRT). It examines how prejudice and intolerance pervade social, cultural, and legal institutions, thus perpetuating systemic inequality based not only on
race but also on gender, class, and sexual orientation. CRT challenges dominant narratives about race and racism in society,
arguing that they are not merely individual acts of prejudice but
are deeply ingrained in the very fabric of society.

CRT holds that race in humans is not a natural, biologically
grounded feature of physically distinct subgroups, such as it
is in animals that are taxonomized into distinct groups on the
basis of inherited physical differences, shared characteristics, and
evolutionary relationships. Race is a socially constructed category, according to proponents of CRT, that is used to oppress

and exploit people of color. Hence race and racism are embedded in social, cultural, and legal institutions, perpetuating systemic inequality. By addressing the ways in which race and racism intersect with other factors, such as gender, class, and sexual orientation, CRT seeks to challenge and transform these structures.

Liberals—again, according to proponents of CRT—do not address the fundamental injustices that minorities face in society. By focusing only on the most obvious racist practices, for example, overt miscarriages of justice, they supposedly refuse to know about other occurrences of bigotry and prejudice. The very emphasis on the equitable treatment of everybody under the law (color blindness)— a laudable liberal objective—allegedly ignores the indirect, subtle, or systemic manifestations of racism.

But then there are critics of CRT. According to them, not everybody, especially not high school students, should learn about such issues. They argue that the theory is divisive, emphasizes racial differences rather than promoting unity, and that it can create a sense of victimhood among minority groups. Parents must have the right to ensure that their children remain ignorant of "woke" ideas and theories.

Hence, in some states, the powers to be decided that CRT is off limits. As an example, let me cite Florida's Individual Freedom Act, commonly known as the *Stop Wrongs to Our Kids and Employees (WOKE) Act*. Signed into law in 2022, House Bill 0007 regulates what may *not* be taught in schools and workplaces. It essentially prohibits teachers to subject (sic!) their students to instruction that may lead them to believe that "a person's moral character or status as either privileged or oppressed is necessarily determined by his or her race, color, national origin, or sex," or that "a person, by virtue of his or her race, color, sex, or national origin, bears personal responsibility for, and must feel guilt, anguish, or other forms of psychological distress because of actions, in which the person played no part, committed in the past by other members of the same race, color, national origin, or sex."

In plain language the act says that teachers may not say that non-white people are being discriminated, or that white people should feel distressed or guilty about what their forefathers did to their non-white countrymen.

People may have differing opinions about the motivations for the act. But one would think, at the very least, that these are subjects worth discussing, especially in a country that proudly proclaimed freedom of speech as the First Amendment to its Constitution. Not so in states that enact laws against even just raising these subjects in school discussions.

The debate around CRT in schools has intensified in recent years. Supporters of the ban argue that CRT is an inappropriate and harmful way to teach about race and that it promotes a

PARENTAL ADVISORY EXPLICIT CONTENT

Parental Advisory label in Florida schools

negative view of American history and institutions. They also claim that CRT is indoctrination rather than education, as it presents a particular perspective as the absolute truth.

Proponents of CRT argue that the theory provides a necessary and more accurate understanding of the historical and ongoing impact of racism in society, and that it fosters critical thinking and dialogue about race and inequality. Banning CRT from schools would stifle important conversations about race, they say, and silence the voices of marginalized communities.

ک

Six days after the Individual Freedom Act was signed, the Florida governor signed House Bill 1557, the Parental Rights in Education bill, aka the "*Don't Say Gay*" bill. It provides that "Classroom instruction by school personnel or third parties on sexual orientation or gender identity may not occur in kindergarten through grade 3 or in a manner that is not age-appropriate or developmentally appropriate for students in accordance with state standards."

One may agree or disagree with the bill. The fact is that Florida insists on keeping children ignorant about questions of sexual orientation and gender identity until age ten.

53

OLDIES PREFER IGNORANCE

Twenty-Somethings vs. Over-Eighties

Many people willfully forego certain information; they prefer to remain ignorant, for example, about the date of their or their partner's deaths, the outcome of a football game, and whether their marriage will end in divorce. People may decide to forego such information when they anticipate that regret over negative information will exceed the relief triggered by learning of a positive outcome. The authors of the 2017 study, Gert Gigerenzer in Germany and Rocio Garcia-Retamero in Spain, named this effect "anticipated regret" (see chapter 50).

Four years later Ralph Hertwig, Jan K. Woike, and Jürgen Schupp in Berlin interviewed more than two thousand people about their predilection for knowledge or ignorance.[1] They categorized their sample according to age so they could investigate whether older people differed from younger people in their preference for ignorance.

They presented thirteen scenarios to the interviewees that were designed to capture regulation of emotions and avoidance of anticipated regret. The situations depicted involved the risk of major losses and negative emotions as well as the possibility of major gains (e.g., relief of emotional strain and existential anxieties). For each scenario, the subjects were asked to decide:

"Would you want to know or would you prefer not to know?" Depending on the scenario, between 21 percent and 90 percent of the interviewees would prefer to remain ignorant (all ages):

Prefer Not to Know (Deliberate Ignorance Scenarios)

Knowledge of one's exact date of death.	90%
Knowledge of a job applicant's gender and appearance.	72%
Knowledge about the past faithfulness of one's future spouse.	56%
Knowledge about the existence of a god-like deity.	54%
Knowledge of a deceased relative's potential Nazi past.	54%
Knowledge of a potential genetic mutation linked with an incurable disease.	53%
Knowledge of colleagues' bonus payments.	51%
Knowledge of having potentially eaten horsemeat (instead of beef).	49%
Knowledge about treatment of a persecuted ethnic minority in a brutal dictatorship.	43%
Knowledge of one's genetic predisposition for various diseases.	42%
Knowledge of whether one's friends or family worked as informants.	38%
Knowledge about the probability of succeeding in a new business.	36%
Knowledge of the authenticity of a recently purchased antique statue.	21%

Especially interesting in the context of Germany are the questions pertaining to a potential Nazi past of family members (54 percent prefer not to know) and to potential informants of the Stasi (*Staatssicherheitsdienst*, the East German State Security

Service). "Suppose you had lived in East Germany as an adult. By consulting your Stasi file, you would be able to find out whether close friends or family members worked as informants and spied on you." In responding to this idea, 38 percent preferred not to know. (The report separated males and females. I report the rounded average.)

The authors next examined deliberate ignorance as a function of age. Does age play a role in decisions whether or not to gain information? As they found, the preference for ignorance, by and large, increases with age. On average, compared to the group of twenty-somethings, 17 percent more people from the group of

Emblem of the Stasi, the East German State Security Service

over-eighty-year-olds preferred to forego information. Although it may be difficult for over-eighty-year-olds to put themselves into the shoes of twenty-somethings about to get married, there were 22 percent more of them who stated that they would prefer to forego information about the past faithfulness of a future spouse. On the other hand, the old and the young would like to remain ignorant about the gender and appearance of an applicant for a position in an orchestra in about equal proportions (70 to 72 percent).

Differences in "Prefer Not to Know" Responses:
Over-Eighty-Year-Olds vs. Twenty-Somethings

Knowledge of a potential genetic mutation linked with an incurable disease. +32%

Knowledge of one's genetic predisposition for various diseases. +29%

Knowledge of a deceased relative's potential Nazi past. +25%

Knowledge about treatment of a persecuted ethnic minority in a brutal dictatorship. +23%

Knowledge about the past faithfulness of one's future spouse. +22%

Knowledge about the probability of succeeding in a new business. +19%

Knowledge of the authenticity of a recently purchased antique statue. +19%

Knowledge of having potentially eaten horsemeat (instead of beef). +19%

Knowledge of whether one's friends or family worked as informants. +17%

Knowledge of one's exact date of death. +11%

Knowledge of colleagues' bonus payments. +6%

Knowledge about the existence of a god-like deity. +5%

Knowledge of a job applicant's gender and appearance. +1%

The result of the study raises the question of why the preference for deliberate ignorance is more pronounced in older age. One explanation the authors proposed is that the future is wide open for young people and any information allows them to prioritize future-oriented goals: extend their social network, expand their intellectual horizon, and organize their lives. Older people, on the other hand, favor present-focused goals, such as emotional meaning and satisfaction. With a smaller window of time left, they want to reduce the risk of potentially negative information, even if it means foregoing potentially positive information. This is especially true for information about health: the shorter the remaining life expectancy, the lower the expected utility of additional knowledge.

&.

Amusingly, the findings give rise to an intriguing paradox: the propensity of twenty-somethings to seek information that the over-eighties would prefer to ignore means that younger adults may acquire knowledge that their older selves would have preferred not to have . . . but by then it's too late.

54

PROTECTING THE SECOND AMENDMENT

The Dickey and Tiahart Amendments

Lamentations about gun violence in the United States have become standard fare for well-meaning progressives and liberals. In 2020, there were 45,222 firearm-related deaths in the United States—that's about 124 people dying each day due to the use of firearms. More than half of these deaths were suicides, and more than 40 percent were homicides. With so many lives lost each year due to gunfire—murders, gang battles, suicides, mass shootings, accidents, shootings by the police—in-depth knowledge about the reasons for the bloodbaths is sorely needed to find ways to prevent, or at least decrease, the human toll. Surprisingly—or maybe not—the powers to be do not want to know.

The Second Amendment gives American citizens the constitutional right to carry firearms. Anything that gets in the way of unlimited gun ownership is anathema to the National Rifle Association (NRA). (In 2021, the NRA attempted to file for bankruptcy, but the application was denied by the court.) Gun manufacturers, gun retailers, and gun enthusiasts represent a formidable lobby in Congress that holds most members of the Republican Party in awe.

"Concealed gun allowed" sign at a gas station in New York

Firearm incidents are a leading cause of premature deaths in the United States, so their occurrence can rightfully be described as an epidemic, which is defined by the *Merriam-Webster Dictionary* as "an outbreak or product of sudden rapid spread, growth, or development." As such, control and prevention of firearm incidents fall within the purview of the Centers for Disease Control and Prevention (CDC), "the nation's leading science-based, data-driven, service organization that protects the public's health." Because "firearm injuries are a serious public health problem," the CDC is the federal agency that is tasked with research on this epidemic.[1] (Note that *Merriam-Webster*'s definition does not limit epidemics to "diseases.")

To gain a true understanding of the public health, financial, psychological, and social toll of gun injuries, many questions require answers. How often is anyone injured by a firearm in America? Why, how, and what kinds of weapons are used? What are the underlying causes? What are the relationships between shooters and victims? To develop and evaluate programs that could help prevent criminal shootings, accidents, and suicides, people in public health, criminal justice, policing, and academia need to acquire data and information.

However, for many decades, the US government put hurdles in the way of relevant research. The NRA, firearm manufacturers, and gun retailers did everything in their power to keep the public ignorant of the true extent of the harmful consequences of gun ownership. The gun lobby feared that academic research could undermine the Second Amendments rights of American citizens to own firearms.

This had not always been so. In 1993, the CDC supported a study that found "keeping a gun in the home was strongly and independently associated with an increased risk of homicide."[2] The study showed that the risk of killing or being killed by a firearm in a household with guns was nearly three times as high as in households without guns. This was information the likes of which the gun enthusiasts did not want. They prefer to point to academic studies of instances where gun ownership possibly prevented a crime.[3]

To avoid further inconvenient research, the NRA lobbied Congress and three years later the Republican congressman James Dickey managed to add an amendment to the 1996 spending bill of the federal government. The Dickey Amendment stated that "none of the funds made available for injury prevention and control at the Centers for Disease Control and Prevention (CDC) may be used to advocate or promote gun control." Funds that the CDC had invested in firearm injury research the previous year were reallocated to a different purpose. Although the amendment did not expressly forbid research on the public health problems due to firearms, it had the effect the NRA desired: research stopped.

Efforts to keep the public in ignorance did not stop with the CDC. Other thorns in the gun lobby's eye were the Bureau of Alcohol, Tobacco, Firearms, and Explosives (ATF) and the FBI. Lobbying efforts, in particular by the NRA's Institute for Legislative Action, again did their part. The excuse for their efforts to suppress data collection and research was that the "NRA is committed to ensuring confidentiality of sensitive law enforcement information."

The Republican congressman Todd Tiahart amended clauses to the 2003 spending bill that prohibited the ATF from releasing firearm trace data and from requiring gun dealers to submit their inventories to law enforcement. The bill also required the FBI to destroy all records of approved gun purchases within twenty-four hours.

Just as the Dickey Amendment put reins on research into the consequences of gun ownership, the Tiahart Amendment limited the efforts to study the trafficking of firearms. Law enforcement's ability to investigate gun crimes and prosecute unscrupulous gun dealers has since been significantly restricted.

Fortunately, with time, saner minds prevailed. Limitations to knowledge and the dissemination of data to keep the public in ignorance started to go out of fashion. In 2018, Congress passed a spending bill that included language restoring the CDC's authority to resume gun-related studies. However, no funds were allocated, so research did not take off. Only in the federal budget of 2020 was money allocated for research on preventing deaths and injuries from guns. Twenty-five million dollars were split equally between the National Institutes of Health and the CDC to fund research that would shed light on the effects of gun ownership.

The NRA was not about to give up. Its flagship publication *American Rifleman* declared that it "fights back against politicians, judges, and bureaucrats who want to regulate, restrict, and ultimately, destroy your Second Amendment freedom." The freedom to know is not guaranteed in the Constitution or in its amendments, so ignorance is king.

55

IGNORING HIGH-RISK EVENTS

Disaster Insurance

I n 1978 a team at the University of Pennsylvania's Wharton School, led by Howard Kunreuther, published a report that raised eyebrows. It was about homeowners' astonishing unwillingness to buy insurance against disasters such as floods and earthquakes. In the preface, Kenneth Arrow, then at Harvard University, called it a "pathbreaking study in opening up a new field of inquiry, the large-scale field study of risk-taking behavior." Arrow, the Nobel Prize winning economist, knew all there was to know about efficient markets and rational individuals. Nevertheless, "even someone like myself . . . has to be surprised at the failure of the flood-insurance market," he wrote in the preface to Kunreuther's study.[1]

The key notion in anything that has to do with insurance is the expected loss. It is defined as the dollar amount of the damages if the disaster occurs, multiplied by the probability that the disaster does, in fact, occur. If a house is valued at $100,000 and there is a one in a thousand chance that it will be totally destroyed in a flood or an earthquake, then the expected loss is $100 (= 0.001 x 100,000). To this expected loss, aka actuarial value, the insurance company adds administrative expenses, a safety margin to cover its own risk, and profits. This is how the premiums the company charges are calculated. So, in total the cost to insure such a house against floods or earthquakes could come to about $150.

According to so-called expected utility theory, rational, risk-averse people should always be willing to insure their home against a disaster if the premium is not too much above the actuarial value. Hence, in this example, if the probability of a disaster is 0.001 per year, the homeowners should be prepared to pay somewhat more than $100, say $150, every year to protect themselves from financial loss. (The insured person will not be totally covered. To counter frauds and frivolous claims, insurance companies usually include a "deductible" that is subtracted from the final payout.)

੨●

In 1968 the National Flood Insurance Program was enacted in the United States as a means of offering federally subsidized flood insurance on a nationwide basis through the cooperation of the federal government and the private insurance industry. In effect, due to the subsidies, the insurance premiums that were offered to homeowners were close to, or even below, actuarial values. (In the above example, that would mean that the insurance premium would be only, say, $90.)

And what did Kunreuther's team find out? According to the theories of Arrow and other economists, rational, risk-averse homeowners should have been "willing to pay something more than an actuarially fair premium to gain protection against loss." But in face-to-face interviews with more than two thousand homeowners in thirteen states, the Wharton team realized that in spite of the subsidies people frequently failed to obtain

Insured against disaster?

insurance for their homes, even when the terms were so favorable that any rational, risk-averse consumer should have purchased insurance. The field study was followed by laboratory experiments that reinforced these findings.

Consumers sometimes make nonoptimal decisions because they have difficulty dealing with low-probability, high-risk events. Instead of maximizing their expected utility, as traditional economists would have them do (traditional in the 1970s, that is), individuals often ignore facts and instead utilize simplified rules of thumb to make decisions. These rules of thumb, commonly called heuristics were investigated not by economists but by psychologists. The gist of the psychologist's new theories was that "rational behavior" must be replaced with behavior that is compatible with human abilities and limitations. Many homeowners do not have access to information, don't know where to get data, cannot process information, and lack the computational abilities.

After all, gathering information and collecting data about the hazard (the probability of a disaster and the expected loss), on one hand, and the insurance options, on the other, entails some effort. Then there are computations to perform. It is usually not as simple as multiplying the value of the house with the probability of a flood or an earthquake. Indeed, the possibility of partial damages must be taken into account, as well as the deductible, and then there's the fine print that sets out all the exclusions. In addition, the insurance company may insist that the homeowner install reinforcements and take other mitigating measures for the policy to be valid, etc., etc.

Of these heuristics (or shortcuts, or rules of thumb), the most important for insurance decisions is that probabilities below a threshold are simply treated as zero. If a disaster happens very rarely, homeowners believe it will never happen to them. Hence, given that so much effort is involved and that the probability is so low anyway, many homeowners decide just to ignore the entire matter. (Trigger warning: I do not condone this type of behavior.)

56

LEGISLATING MATHEMATICAL TRUTH

π = 3.2?

everybody beyond middle school knows (or should know) that the circumference of a circle is equal to the number π, multiplied by the circle's diameter, and that the area of a circle is π, multiplied by the square of the circle's radius.

The importance of the number π goes way beyond geometry; in fact, it is ubiquitous in mathematics, physics, statistics, all fields of science, construction, and engineering. Today it is known that the numerical value of π is 3.141592653589793238 . . . with a never-ending string of decimals following. But for a very long time this was unknown.

❧

That the ratio of a circle's circumference to its diameter is about three was recognized by ancient civilizations four millennia ago. A Babylonian tablet (ca. seventeenth century BCE) put the value of that number at 3⅛. A couple of centuries later the Rhind Papyrus, an Egyptian mathematical treatise (approximately 1550 BCE), stated that the ratio of a circle's circumference to its diameter is ²⁵⁶/₈₁, which corresponds to about 3.16049383. . . . In the third century BCE, Archimedes, a Greek mathematician, did

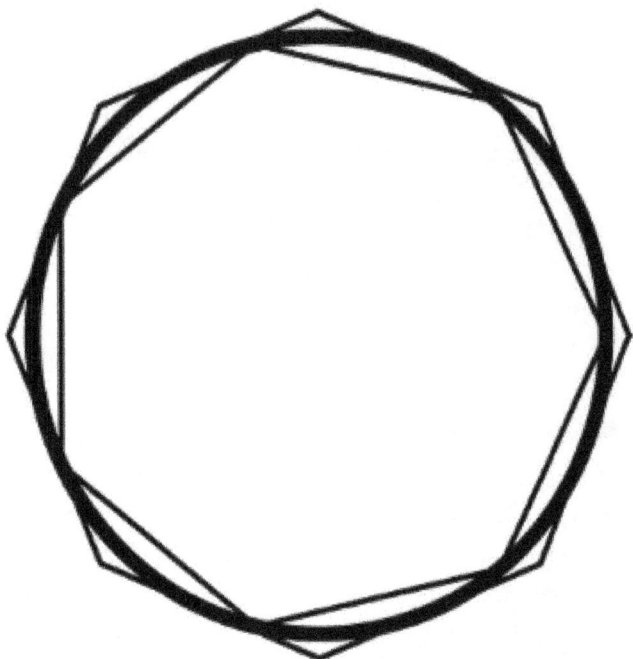

Archimedes's method of approaching the area of a circle. Image by author.

even better. He approximated the value of π by inscribing and circumscribing the circle with polygons. He stated that because the circle's area is a bit larger than the area of the inscribed polygon, and a bit less than the area of the circumscribed polygon, the area of the circle and hence the value of π can be approximated ever more exactly by increasing the number of sides of the polygons. Using this algorithm (although it was not called an algorithm at the time), Archimedes showed that π must lie between $3^{10}/71$ and $3\frac{1}{7}$, i.e., between 3.140845 . . . and 3.142857 . . . (see also chapter 16).

But the numerical value of π cannot be expressed exactly. In the 1760s, the Swiss mathematician Johann Lambert proved that π is irrational: it cannot be expressed as a ratio of integers. In other words, the decimal expansion never ends and never repeats. (In contrast, the ratio ½ equals 0.5 and ends there; the ratio ⅐ equals 0.1428571428577142857 . . . and keeps repeating 142857.)

Apparently, the news of Lambert's proof traveled very slowly and had not yet reached the State of Indiana at the end of the nineteenth century. There, a country doctor by the name of Edward Goodwin, practicing medicine in Posey County, "was supernaturally taught the exact measure of the circle." The revelation, which he ascribed to scriptural statements and promises, was that the ratio of the diameter of a circle and its circumference "is as five-fourths to four," i.e., as 4 divided by 1.25. This equals the nice round number 3.2.[1]

Goodwin's revelation was based on mathematical "facts" that contradicted elementary geometry. For example, he claimed that the diagonal of a square with side-lengths 7.0 was 10.0. This is, of course, wrong: by Pythagoras's Theorem, the correct length of the diagonal is 9.8994. . . . Based on such false premises, Goodwin arrived at his "new mathematical truth," which claimed that π equals 3.2. As an approximation this is not too bad; but as a mathematical truth it is completely and utterly wrong.

Goodwin was ignorant of everything mathematicians had found out about π during the past several millennia. Hoping to strike it rich, he copyrighted his revelation in the United States and in several countries in Europe. As a true patriot, however, he was prepared to offer it without compensation to his compatriots in the State of Indiana as long as his mathematical truth was "accepted and adopted by the official action of the Legislature of 1897."

To justify the proposed bill, its Section 3 cites the fact that "his solutions of the trisection of the angle, duplication of the cube and quadrature of the circle having been already accepted as contributions to science by the *American Mathematical Monthly*, the leading exponent of mathematical thought in this country." To further underline how important Goodwin's revelation was, the bill ends with these words: "And be it remembered that these noted problems had been long since given up by scientific bodies as insolvable mysteries and above man's ability to comprehend."

It is true that the three issues—trisecting any angle, duplicating a cube, and squaring a circle, using only compass and ruler—are problems that are unsolvable because they all require the construction of irrational numbers. With ruler and compass this was, is, and always will be impossible. Hence it is obviously not true that Goodwin found solutions to these problems. It is true, however, that Goodwin's nonsensical fantasy was published in the *Mathematical Monthly*. But it is definitely not true that his paper was accepted as a contribution to science. In fact, the byline stated that the paper was "published by the request of the author;" hence it was more an infomercial than a mathematical paper, and an incoherent, erroneous, nonsensical one at that.

Nevertheless, the House dutifully referred Bill 246 "for an act introducing a new mathematical truth and offered as a contribution to education to be used only by the State of Indiana free of cost," to the Committee on Canals that, in turn, referred it to the Committee on Education. It was debated in the full House on February 5, 1897, and passed unanimously, 67 to 0.

By fortuitous coincidence, on that very day a professor of mathematics and Latin, Clarence Abiathar Waldo of Purdue University, happened to be present in the Statehouse. He was there to lobby for his university's budget appropriation, but when he heard that a mathematical bill was being debated in the

General Assembly, he wanted to listen in . . . and was horrified. He decided to coach the senators who were to debate the bill a week later about its absurdity. In the meantime, newspapers, local as well as out-of-state, had picked up the story and had a field day ridiculing it.

Still, the bill was received favorably in the Senate's Temperance Committee and passed on for debate to the Senate floor. By now the senators, coached by Waldo and derided far and wide, had become wary of the bill. "The Senate might as well try to legislate water to run uphill as to establish mathematical truth by law," one of them remarked, adding that the Indiana State Legislature had laid itself open to ridicule by leading newspapers of Chicago and the East.

During further debate, the senators admitted that they were quite ignorant of the merits (and obviously also of the demerits) of the proposition. They decided that consideration of such a proposition was neither dignified nor worthy of the Senate's consideration, and that mathematical truths were not a subject for legislation. To put an end to the debate, an astute senator moved to postpone the bill indefinitely. The motion carried . . . and that is where the matter stands until this day.

57

TO TEST OR NOT TO TEST?

Prostate-Specific Antigens

When a man reaches the age of about forty, he is often encouraged to have his blood tested for the presence of prostate-specific antigens (PSA). For once, I can forego in this chapter use of gender-nonspecific pronouns. No he/she or, *horribile dictu*, "them" when I mean "him." It will be "he" throughout, and "they" only if I mean two men or more.

The PSA is a protein produced by normal as well as by malignant cells of the prostate gland. If the prostate gland is enlarged, more PSA is produced. The crucial question is: Is the prostate gland enlarged due to normal causes, such as the man's age (benign prostatic hyperplasia), or due to prostate cancer (malignant prostatic hyperplasia)? Or is there a totally unrelated cause (inflammation of the prostate, vigorous exercise, recent sexual activity, horseback riding, or bicycling)?

The test measures the level of PSA in the blood. If there are less than four nanograms of PSA per milliliter of blood, the test is said to show no indication of prostate cancer. If there are more than ten nanograms per milliliter, it is an indication that the testee may have cancer. The range between four and ten nanograms is a gray zone: he may or may not have prostate cancer; one does not know. But there's a lot more that one does not know, as you shall see.

After the discovery of PSA in the 1970s as a clinically useful biomarker for prostate cancer, it seemed a good idea to screen all men approaching middle age. After all, because the first stages of prostate cancer occur without symptoms, early detection could significantly reduce mortality. Whenever PSA was high, treatment would be recommended. In the 1990s, screening took off.

But then a number of professional medical organizations began to have second thoughts. The reason is that once one obtains the test results, there is good news and bad news . . . whatever the result. If the level is above ten ng/ml—in itself a concerning result—there is a good chance that it is a "false positive," i.e., the high levels of protein were produced by the benign cells of the prostate gland. However, a level of below four nb/ml does not guarantee that the man is cancer free; the test result could indicate a "false negative," i.e., only a low level of the protein was detected even though the prostate is enlarged due to cancer.

So what knowledge has one gained?

A study by the US Preventive Services Task Force estimated that of a thousand men between the ages of fifty-five and sixty-nine who are screened regularly over a period of thirteen years, 240 would have a positive PSA result. But after additional testing, for example, with a biopsy, only one hundred would actually be diagnosed with prostate cancer.[1] This translates to about 60 percent false positives. Of the one hundred correct positives, many would have survived their cancer without treatment, and some would have died even with treatment. Altogether the task force estimated that about one to two deaths could be avoided over a period of thirteen years by regular screening of one thousand men. But for each man saved from early death, there would be about 100 to 150 men who would undergo superfluous surgery or radiation treatment (with all the risks and frequent side effects) and needless worry.

Why did the study only consider men younger than seventy? Well, prostate cancer develops very slowly. Most prostate cancers would never progress to a clinically meaningful stage if left undiagnosed and untreated during a man's lifetime. Even if prostate cancer is identified, the ten-year survival rate is about 98 percent, and the fifteen-year survival rate is about 95 percent: this means that ten years after diagnosis, the average prostate cancer patient is just 2 percent less likely to survive than a man without prostate cancer, and only 5 precent less likely to survive fifteen years. Therefore, men over the age of about seventy, who do not expect to live much longer than for another ten or fifteen years anyway, are excluded from many studies.

A study of the National Cancer Institute found that men who underwent annual PSA screening did show a higher incidence of prostate cancer than men in the control group, but, significantly, they had about the same rate of deaths from the disease.[2] Hence, many men were treated for prostate cancers that would not have been detected in their lifetime without screening. Consequently, these men were exposed unnecessarily to the potential harms of treatment.

A study in Europe concluded that forty-eight men would have to be treated to save one life. Many of the forty-seven who were unnecessarily treated would suffer from side effects for the rest of their lives.

As a consequence of similar studies, many medical organizations reevaluated their early recommendations. In 2015 the United States Preventive Services Task Force recommended against PSA-based screening, because the potential benefits of reducing mortality probably didn't outweigh the risks of overdiagnosis and overtreatment. Other countries follow different protocols.

Three decades after PSA screening was introduced, the consensus seems to be that men over seventy should not get tested at all. As for younger men, although PSA screening does show a minute reduction of mortality, men between fifty-five and sixty-nine years of age might be better off remaining ignorant of their PSA levels.

58

BEWARE: VIOLENCE AHEAD!

Trigger Warning

It has become fashionable at various universities lately to "prepend" trigger warnings before assigning, displaying, or discussing reading material, films, or artworks in classroom settings. Designed to spare students from distress and suffering, these warnings advise emotional students to be aware of what awaits them and maybe to skip the assignment.

The practice is quite controversial. On one hand, it is certainly laudable to spare people who have been traumatized in the past—combat veterans, victims of sex crimes, survivors of child abuse—from reviving distressing experiences in their minds and suffering further anguish. To warn sufferers from epilepsy that a performance they are about to see contains certain visual and auditory stimuli—flashing beams, rapid changes in lighting, sudden noises—is indispensable because such stimuli may potentially trigger an epileptic episode. And to keep children from watching films that contain sex and violence, the PG-13 and R movie rating is certainly warranted.

On the other hand, many subjects that are part of human history and of today's society are upsetting or offensive: race, sexual orientation, disability, colonialism, torture. People can be of different minds about whether one may belittle religious sensitivities, for example, by exhibiting cartoons of the prophet Muhammad in a classroom setting. But alleged sensitivities go much further. Shakespeare's Romeo and Juliet may provoke suicidal thoughts, Michelangelo's David shows a naked man, blockbuster movies such as *To Kill a Mockingbird* address themes of racism, discrimination, and violence, *Schindler's List* includes scenes of violence, genocide, and discrimination during the Holocaust, and *A Clockwork Orange* contains graphic depictions of violence, sexual assault, and societal dystopia. Famous paintings such as *The Scream* by Edvard Munch are meant to evoke feelings of anxiety or unease, and the *Guernica* by Pablo Picasso depicts unsettling imagery about the horrors of war. Botticelli's *Birth of Venus* contains nudity and may be considered provocative or sexually suggestive.

Historical novels such as *Uncle Tom's Cabin* describe the horrors of slavery, and Remarque's *Nothing New in the West* depicts the horrors of war. *Lolita* by Vladimir Nabokov explores themes of pedophilia, Thomas Salinger's *The Catcher in the Rye* explores themes of mental health, isolation, and teenage angst, and *The Handmaid's Tale* by Margaret Atwood includes themes of oppression, sexual assault, and dystopian society.

It was Picasso's stated aim to evoke horror in the viewers, and it was Remarque's specific desire to perturb the readers. Should sensitive students be kept in ignorance of works of art and literature because they trigger uncomfortable thoughts? Is it not the mission of an institute of learning to challenge conventional thinking and preconceptions? Opponents of trigger warnings maintain that students should get no dispensation from dealing with subjects that disturb them. Instead of coddling future leaders of society, industry, and politics, and shielding them from uncomfortable theories, students should be confronted with the very ideas that contradict their biases and prejudices.

This is a work of fiction.

CONSIDER
YOURSELF
TRIGGER WARNED

Trigger warning

Protecting students, rather than challenging them, is infantilizing and anti-intellectual, many educators believe. Where if not in the classroom would discussions expand the minds of young citizens? According to opponents of trigger warnings, intellectual comfort should not come before intellectual engagement. In a report on Academic Freedom and Tenure, the American Association of University Professors wrote that "some discomfort is inevitable in classrooms if the goal is to expose students to new ideas, have them question beliefs they have taken for granted, grapple with ethical problems they have never considered, and, more generally, expand their horizons so as to become informed and responsible democratic citizens."[1]

And the University of Chicago expresses it thus in a letter to incoming students: "Our commitment to academic freedom means that we do not support so-called 'trigger warnings,' . . . and we do not condone the creation of intellectual 'safe spaces' where individuals can retreat from ideas and perspectives at odds with their own."[2]

≈

Trigger warnings may have the opposite effect of what they are meant to achieve, namely, to protect the vulnerable in a meaningful way. Trigger warnings might inadvertently reinforce stereotypes or assumptions about individuals with post-traumatic stress syndrome (PTSD), framing them as fragile or incapable of engaging with challenging content. This can perpetuate stigma. In addition, some psychologists maintain that shielding individuals from potential triggers might actually hinder their progress in developing coping mechanisms, resilience, and post-traumatic growth; it impedes healing and growth in individuals with PTSD. In any case, they maintain, classrooms are not the venue to treat PTSD.

Proponents of trigger warnings argue that emotional well-being is as important as intellectual pursuit and should be considered alongside academic training. By providing advance notice of potentially distressing content, individuals can make informed decisions about their engagement with potentially painful themes and take necessary steps to manage their mental health. Trigger warnings are not meant to let students remain ignorant by avoiding confrontation with contentious subjects, proponents say, but to allow them to prepare themselves mentally for engagement with painful topics at their own pace, whenever they feel ready. Forewarned students are able to summon coping strategies and thus be better equipped to approach distressing subjects.

On top of that, warnings may convey valuable information to people unconcerned with the trigger. By making the general public aware of potential biases, stereotypes, or sensitive historical events that may affect people from marginalized communities, the trigger warnings foster a more inclusive and respectful learning environment. They not only protect the concerned but promote respect, empathy, and consideration by the unconcerned.

Judge for yourself!

59

DON'T TELL ME YOU DID IT

Plausible Deniability

Quentin, a defense attorney, has been hired by Bruce, a member of a street gang, to defend him for charges of manslaughter. To be represented effectively in court by his attorney, Bruce must be completely frank with Quentin and tell him everything. Only if the defense attorney is aware of all pertinent facts can he represent his client. But there are circumstances when a lawyer might wish to remain ignorant. He might prefer not to know certain details about his client or of a situation. Hence, surprisingly, there may be circumstances in which Quentin might say to Bruce: "Don't tell me! I don't want to know."

Prosecutors and defense attorneys are obliged to be truthful. They are members of the court and have a duty to promote justice and the effective operation of the judicial system. After all, it is neither the prosecutor's role to have a suspect convicted at all costs, nor the defense attorney's role to prove innocence by any means. Both are obliged to help the court determine the truth and to uphold the integrity of the justice system.

Hence defense attorneys ensure that the accused receives a fair trial, that his rights are protected, and that the state meets its burden of proving guilt. They are obliged to advocate for their clients and try to get the best deal for them, even if they're guilty, but they are also bound by ethical rules. Knowingly making false statements, providing false evidence, obtaining information unlawfully, or concealing facts is illegal, even if it would aid the accused in getting off the hook.

Member of the court

So, if Bruce denies the charges but in the course of interviews with Quentin starts to open up and begins to say "I actually did . . . ," Quentin might stop him before he says another word; he does not want Bruce to confess to him. All he needs to know is what the prosecutor knows. Hence he might tell Bruce: "Don't tell me if you did it. Just tell me what the police know." As long as Quentin is not actually aware of whether Bruce is guilty—even if he might suspect that he is—he can represent Bruce with vigor.

If a client nevertheless admits guilt to his attorney but wants to claim innocence in court, the attorney is placed in an ethical dilemma. In such a situation, the attorney can defend the client by challenging the prosecution's evidence, examining the legality of procedures such as searches and arrests, and arguing for the client's rights . . . without actually asserting his factual innocence.

&

It does not even matter whether Quentin knows or believes that Bruce is guilty or innocent; it is irrelevant to the case. There is a difference between factual guilt and legal guilt. Factual guilt is what the defendant actually did, and legal guilt is what a prosecutor can prove. An accused may be factually guilty (and the defense attorney may know it), but if the prosecutor does not have enough evidence to prove it, the accused is legally innocent.

Therefore, because the defense attorney is forbidden from telling untruths, he may be better off not knowing everything. If the accused then conceals evidence or perjures himself during the trial, he can honestly claim that he himself acted in good faith. Because he was ignorant of the specific facts and events, he cannot be accused of having encouraged such illegal behavior.

If in spite of the attorney's precaution, the accused has admitted to him that he is guilty but insists that he, nevertheless, intends to testify falsely, the attorney may, as a last resort, ask the judge to let him withdraw from the case; but the court's permission to do so will be required.

An accused must also know that although attorney-client privilege is a basic principle in the legal profession—nothing the client shares with his attorney may be divulged without his consent—it is not sacrosanct; there are loopholes. If the attorney becomes aware, for example, that the client plans to commit a crime, especially one that could result in harm to others, this attorney must do all he can to prevent it. If that means disclosing this information to the court or to law enforcement, so be it, even at the risk of being disbarred. So it is better not to know about it.

ॐ

Remaining ignorant of certain details is also recommended under certain conditions in commercial law. If two lawyers of the same law firm represent two clients who are structuring a business deal together, or who have a dispute with each other, it may be appropriate for the attorneys to remain ignorant of the other side's position. An attorney who moves to a new firm might need to recuse herself from certain meetings to remain ignorant of cases handled by the new firm that could present conflicts with clients from the previous firm.

An attorney representing multiple clients might prefer to remain ignorant of one client's dealings if they overlap with a second client's interests, or if knowing details of a present case could reveal a conflict with another client or a past case. Sometimes too much knowledge might bar an attorney from representing a client.

60

EXPERTS, SHMEXPERTS

Denial of Expertise, Dismissal of Elites

In times gone by, people tended to trust experts, be they doctors, engineers, scientists, teachers, lawyers . . . even priests and politicians. You name the area of expertise, and there were authoritative professionals for it. These were the intellectuals, the nation's elite, who did the thinking and deciding for the "common man." Met with deference, their opinions and judgments were respected and mostly followed.

No more; today it has become fashionable to distrust experts. Professionals are ignored, institutions are distrusted, and intellectuals are snubbed. Opinions by experts and assertions by professionals are nothing but fallacious "appeals to authority," sure signs of dreadful "elitism," and an obvious effort to use credentials to stifle the dialogue required by a "real" democracy. And for the common *Homo sapiens* to refuse to acknowledge alternative facts, no matter how fantastic or inane, is to be closed-minded.[1]

Although unpretentious people used to be aware of their ignorance, their descendants today (namely, we) consider themselves (ourselves) knowledgeable on just about anything that hits their (our) fancy. With the ubiquitous availability of information and misinformation on the internet, truths and half-truths in the printed press, and news and fake news on TV, everybody may believe that he or she can be a true expert on medicine, science, economics, foreign policy, governmental administration, culture, history, and so on.

How did distrust in traditional experts arise? Religious authorities used to (and some still do) see intellectual inquiry as a threat to established dogma. Galileo Galilei, the foremost astronomer of his time, had to renounce "the false opinion" that Earth orbits the Sun because the church fathers, experts on anything under the sun, were of a different opinion (see also chapter 47).

Confirmation bias, a psychological phenomenon, also contributes to denial of expert opinions. People seek out information that aligns with their preexisting beliefs, and they tend to ignore information that contradicts them. If a person thinks the Earth is flat, he will remain resistant to expert opinions that challenge that view (see chapter 51). If someone believes that aliens landed on Earth, no astrophysicist will be able to change her belief (see chapter 1). And if a mother is convinced that inoculations against whooping cough cause autism, no pediatrician can persuade her to vaccinate her children.

Commercial interests and the lust for power are other reasons for distrust. Industries threatened by scientific research may choose to cast doubt on that research. Big Pharma, Big Tobacco, Big Oil, and Big Coal made it their mission at various times to cause distrust in science. Populist leaders thrive on the idea that the established system (including its experts) is broken and needs to be overhauled. With information overload, political polarization, confirmation bias, commercial interests, and populism, it is no wonder that the authorities, the professionals, and traditional experts are ignored.

Another reason for the dismissal of expertise may arise from a frontier mentality that prizes practical skills, instinct, and common sense over formal education, book learning, and expertise. After all, education and scientific understanding are hard to come by, and it may require some effort to distinguish between credible sources and noncredible ones. To master a subject requires curiosity and—I dare say—a certain amount of intelligence. Scientific truths, engineering details, and political facts are often difficult to comprehend. Intellectually challenged contemporaries often prefer to dismiss anything they don't immediately understand rather than make the effort to learn something new. After all, who is this professor to tell me what is true? My opinion is as good as hers.

The belief that common men and women know best is reinforced by the idea at the very heart of democracy. Its premise, that every citizen recognizes what is best for him or her, gave citizens a feeling of self-importance and the false conviction that their opinions are as worthy of consideration as that of any expert. This is why Plato distrusted democracy. Aware of the danger of letting common, uninformed, and easily manipulated people make decisions, he believed that a society should be ruled by philosopher-kings who possess the required knowledge and wisdom, i.e., by experts.

ટ️

Doubters like to point to failures of professionals in past situations—mistakes in journalistic reporting or scandals at universities and research centers—as proof that experts and institutions cannot be trusted. Never mind that such failures are relatively rare, that by and large experts have a much better grasp of true facts, and that errors are usually corrected sooner or later. Failures serve to erode trust and public confidence in the intellectual and economic elites of society.

And that brings us to the recurrent distrust of the so-called elites. In the Roman Republic, plebeians sought protection from patrician elites in the Conflict of the Orders; in fourteenth-century England, during the Peasants' Revolt, commoners rose against perceived injustices by the elite; the Protestant Reformation in the sixteenth century was a reaction against perceived corruption and excesses of the Catholic Church's elite; the American and the French revolutions were rebellions against monarchies and aristocracies; and the Russian revolution was against the bourgeois elites.

Memorial to the Peasants' Revolt in the United Kingdom in 1381

The word *elite* derives from eligere, Latin for to choose, to elect. And that says it all because to be chosen is something to which most people aspire. In between revolutions, elites were revered and seen as natural leaders and contributors to culture. Today we admire elite athletes; we rely on elite research institutes; we want our children to be accepted by elite universities; and most economy passengers wish they had elite status so they can be upgraded to first class. They are the chosen . . . and that creates resentment by the nonchosen. So, as the saying goes, if you can't join them, fight them. (Yes, yes, do-gooders claim that the saying is the other way around. But why should their opinion be superior to mine?)

᠁

Although doubters distrust generally accepted expertise and dismiss opinions of acknowledged professionals, they have no qualms about blindly trusting snake oil salespersons, dubious websites, or conspiracy mongerers. Unwilling to learn anything new that could possibly change their minds, many ignorant contemporaries think of themselves as the real experts. This creates a phenomenon that is even worse than dismissal of expertise, namely, the broadcasting of their supposed knowledge far and wide.

Yes, everybody has the right to free speech, everyone has a right to participate in the public sphere—but not on equal terms. The right to free speech does not mean that every contribution to a discussion is worthy of consideration. Reporting on facts, be it in science, politics, or economics, is not a democratic process. Scientific truths are not decided by a majority vote or by whoever shouts loudest. Society needs experts as gatekeepers. Ignorance must not be king.

EPILOGUE

There are many reasons for not knowing things, phenomena, or concepts. In this book I have explored sixty instances of ignorance, analyzing them from four different angles: what we don't know, cannot know, must not know, or refuse to know. But the borders are fluid and often overlap. In this epilogue I highlight a few examples.

Sometimes we ignore truths because problems are unsolved and still await a solution (for example, the Riemann Hypothesis, P vs. NP); such ignorance may be temporary based on the assumption that the human intellect will be capable one day of finding answers. Some knowledge is intentionally kept from us by higher authority, through government order, legal constraints, or institutional secrecy (such as classified documents, restricted scientific research, or proprietary knowledge). Other instances of ignorance arise when we fail to examine our assumptions or biases critically (such as the Dunning-Kruger effect or the Ellsberg Paradox). At times, scientists have simply not looked hard enough (consider dark matter and uncharted depths).

Many questions are unanswerable and will forever remain so because there cannot be a definite and correct answer. We will never know some "truths" because such truths simply do not exist (such as Heisenberg's Uncertainty Principle, the length of the coast of Britain, chaos theory, or KAM theory).

Knowledge may be withheld out of compassion (Professor Bernhardi), to ensure social harmony (compensation secrecy), to maintain impartiality (Rawls's theory of justice), or to fulfill professional obligations (plausible deniability). To remain ignorant often prevents cognitive dissonance by allowing individuals to reject ideas or evidence that conflict with their existing beliefs (such as the flat Earth, the square root of 2, or critical race theory).

An intriguing aspect of ignorance is when people deliberately choose to remain ignorant. This seems quite irrational, but it is only irrational if truth is the sole aim. If personal welfare is considered, it may be quite rational to refuse to know certain truths (such as prostate testing). Occasionally we choose to remain ignorant because the truth would be too difficult to bear (such as testing for Huntington's disease). Disregarding truths may also be the rational course of action if knowledge distresses (trigger warnings), diminishes utility (synthetic diamonds), is difficult to comprehend (the Earth moving around the Sun, π being an irrational number), or agitates too much (older people prefer ignorance). An often overlooked aspect of ignorance is that ignoring certain truths may actually keep decision-makers from making wrong decisions (such as the sunk cost fallacy or irrelevant alternatives).

Ultimately, ignorance in all its forms defines the boundaries of our knowledge. Although in the marketplace of ideas knowledge is usually considered a good, ignorance need not be a bad. At times, it is ignorance that shields us, safeguards our society, preserves our peace of mind, and enhances our decisions.

NOTES

INTRODUCTION

1. See, for example, Matthias Gross and Linsey McGoey, eds., *Routledge International Handbook of Ignorance Studies*, 2nd ed. (Taylor and Francis, 2023).

2. KNOWLEDGE UNDER SEAL: *LES PLIS CACHETÉS*

1. United States Secret Service, Reasons Access May Be Denied Under the FOIA, accessed August 21, 2025, https://www.secretservice.gov/foia /exemptions.
2. See George G. Szpiro, *Pricing the Future: Finance, Physics, and the 300-Year Journey to the Black-Scholes Equation*, chap. 12 (Basic Books, 2011).

3. A MATHEMATICAL RIDDLE: THE RIEMANN HYPOTHESIS

1. Karl Sabbagh, *The Riemann Hypothesis: The Greatest Unsolved Problem in Mathematics* (Farrar, Straus and Giroux, 2003).

4. RESTRICTED ACCESS:
THE FREEDOM OF INFORMATION ACT

1. National Archives, *Code of Federal Regulations*, title 22, chap. 1, sub-chapter A, part 9, section 9.4, https://www.ecfr.gov/current/title-22/chapter-I/subchapter-A/part-9/section-9.4.

2. *Annual Report on Security Clearance Determinations* (National Counter-intelligence and Security Center, April 2020).

5. *IGNORAMUS ET IGNORABIMUS*:
A CONTROVERSY IN GERMANY

1. Estelle du Bois-Reymond, ed., *Reden von Emil du Bois Reymond*, vol. I (Verlag von Veit, 1912), 441–73.

2. David Hilbert, "Mathematische Probleme," *Nachrichten von der Gesell-schaft der Wissenschaften zu Göttingen* (Mathematisch-Physikalische Klasse, 1900), 253–97.

3. Albert Einstein, Sigmund Freud, et al. "Notes and News." *Journal of Philosophy, Psychology and Scientific Methods* 9, no. 15 (July 18, 1912), 419–20.

6. WILL THE UNIVERSE
EXPAND FOREVER?

1. Georges Lemaître, *The gravitational field in a fluid sphere of uniform invariant density according, to the theory of relativity; Note on de Sitter Universe; Note on the theory of pulsating stars* (Massachusetts Institute of Technology, Department of Physics, 1927).

2. Georges Lemaître, "Un Univers homogène de masse constante et de rayon croissant rendant compte de la vitesse radiale des nébuleuses extra-galactiques," *Annales de la Société Scientifique de Bruxelles* (April 1927): 47–59, 49.

7. *IGNORANTIA LEGIS NON EXCUSAT*:
PRESUMED TO KNOW

1. *Commentaries on the Laws of England* (1765–1770), book 4, chap. 2, 27.

2. Robert Joseph Pothier, *Traité des Obligations* (Debure, 1761), translated by William David Evans as *A Treatise on the Law of Obligations or Contracts* (London, 1806.)

3. Oliver Wendell Holmes Jr., *The Common Law* (Little, Brown, 1881).

4. Lambert v. California, 355 US Supreme Court 225 (1957), https://supreme.justia.com/cases/federal/us/355/225/.

8. WHAT IS NOTHING? *HORROR VACUI*

1. Martin Heidegger, *Was ist Metaphysik?* (1929; repr. Verlag Vittorio Klostermann, 2006).

2. Jean-Paul Sartre, *L'être et le néant* (Gallimar, 1943).

9. IS JUSTIFIED TRUE BELIEF KNOWLEDGE? GETTIER'S PROBLEM

1. Plato, *Theaetetus*, in *Plato: Complete Works*, ed. John M. Cooper (Hackett, 1997), 201.

2. Edmund J. Gettier, "Is Justified True Belief Knowledge?," *Analysis* 23, no. 6 (1963): 121–23.

10. HARD BUT EASY: *P* VS. *NP*

1. Manindra Agrawal, Neeraj Kayal, and Nitin Saxena, "PRIMES Is in P," *Annals of Mathematics* 160 (2004), 781–93.

11. IGNORING ONE'S OWN ASSESSMENT: ELLSBERG'S PARADOX

1. Daniel Ellsberg, "Risk, Ambiguity, and the Savage Axioms," *Quarterly Journal of Economics* 75, no. 4 (November 1961), 643–69.

2. Ellsberg, "Risk, Ambiguity, and the Savage Axioms," 655.

12. IT'S THERE BUT CAN'T BE SEEN: DARK MATTER

1. Fritz Zwicky, "Die Rotverschiebung von extragalaktischen Nebeln," *Helvetica Physica Acta*, 6 (1933), 110–27.

2. Vera Rubin and W. Kent Ford Jr., "Rotation of the Andromeda Nebula from a Spectroscopic Survey of Emission Regions," *Astrophysical Journal* 159 (1970), 379ff.

13. UNSKILLED AND UNAWARE OF IT: THE DUNNING-KRUGER EFFECT

1. Justin Kruger and David Dunning, "Unskilled and Unaware of It: How Difficulties in Recognizing One's Own Incompetence Lead to Inflated Self-Assessments," *Journal of Personality and Social Psychology* 77, no. 6 (1999), 1121–34.

16. ON LEARNED IGNORANCE: CARDINAL NICOLAS DE CUSA

1. For more on Cusanus, see George G. Szpiro, *Numbers Rule: The Vexing Mathematics of Democracy, from Plato to the Present* (Princeton University Press, 2010).

17. WHERE AND HOW FAST? HEISENBERG'S UNCERTAINTY PRINCIPLE

1. Werner Heisenberg, "Über den anschaulichen Inhalt der quantentheoretischen Kinematik and Mechanik," *Zeitschrift für Physik* 43 (1927): 172–98.

18. HALF ZEROS, HALF ONES: RANDOM NUMBERS

1. Alan M. Ferrenberg, D. P. Landau, and Y. Joanna Wong, "Monte Carlo Simulations: Hidden Errors from 'Good' Random Number Generators," *Physical Review Letters* 69, no. 23 (1992): 3382–84.

19. SATISFICE, DON'T OPTIMIZE: BOUNDED RATIONALITY

1. Herbert Simon, "Rational Choice and the Structure of the Environment," *Psychological Review* 63, no. 2 (1956): 129–38.

2. Daniel Kahneman and Amos Tversky, "Prospect Theory: An Analysis of Decision Under Risk," *Econometrica*, 47, no. 2 (1979): 263–92.

20. DON'T EVEN ASK: MENO'S PARADOX

1. This chapter is based on chapter 38 in George G. Szpiro, *Perplexing Paradoxes: Unraveling Enigmas in the World Around Us* (Columbia University Press, 2024).

2. George G. Szpiro, *Kepler's Conjecture: How Some of the Greatest Minds in History Helped Solve One of the Oldest Math Problems in the World* (Basic Books, 2003).

23. THE SIXTH SENSE IN A POST-TRUTH WORLD: ALTERNATIVE FACTS AND FAKE NEWS

1. Kellyanne Conway on NBC's *Meet the Press*, January 22, 2017.

2. Dan Mangan, "Trump Told Lesley Stahl He Bashes Press so 'No One Will Believe' Negative Stories about Him," *CNBC*, May 22, 2018.

24. HOW LONG IS THE COAST OF BRITAIN? FRACTAL DIMENSIONS

1. Benoît Mandelbrot, "How Long Is the Coast of Britain? Statistical Self-Similarity and Fractional Dimension," *Science* 156 (1967): 636–38.

2. World Resource Institute, Coastal and Marine Ecosystems—Marine Jurisdictions: Coastline length, Units: Kilometers, https://web.archive.org/web/20120419075053/http://earthtrends.wri.org/text/coastal-marine/variable-61.html.

3. https://users.math.yale.edu/mandelbrot/web_pdfs/howLongIsTheCoastOfBritain.pdf.

25. IS THE SOLAR SYSTEM STABLE? KAM THEORY

1. Jürgen Moser, "On Invariant Curves of Area-Preserving Mappings of an Annulus," *Nachrichten der Akademie der Wissenschaften, Mathematisch-Physikalische Klasse* II (1962): 1–20.

26. ENTERING INFINITE LOOPS: THE TURING HALTING PROBLEM

1. Alan Mathison Turing, "On Computable Numbers, with an Application to the Entscheidungsproblem," *Proceedings of the London Mathematical Society* 58 (1936): 230–65.

27. TRUE BUT NOT PROVABLE: GÖDEL'S INCOMPLETENESS THEOREM

1. Kurt Gödel, "Über formal unentscheidbare Sätze der Principia Mathematica und verwandter Systeme," *Monatshefte für Mathematik und Physik* 38 (1931): 173–98.

28. STORING MUSIC, PHOTOS, VIDEO, AND TEXT: ALGORITHMIC COMPLEXITY

1. Andrey Kolmogorov, "On Tables of Random Numbers," *Sankhyā: Indian Journal of Statistics, Series A*, 25, no. 4 (1963): 369–75.
2. Bertrand Russell, "Mathematical Logic as Based on the Theory of Types," *American Journal of Mathematics* 30, no. 3 (1908): 222–62.

30. IS YOUR RED MY BLUE? QUALIA

1. Frank Jackson, "Epiphenomenal Qualia," *Philosophical Quarterly* 32 (1982): 127–36.

31. PREVENTING LAST RITES: PROFESSOR BERNHARDI

1. Clarence H. Braddock III, "Truth-Telling and Withholding Information" (Department of Bioethics and Humanities, University of Washington, 2018), https://depts.washington.edu/bhdept/ethics-medicine/bioethics-topics/detail/82.
2. Marjorie D. Wenrich et al., "Communicating With Dying Patients Within the Spectrum of Medical Care from Terminal Diagnosis to Death," *Archives of Internal Medicine* 161, no. 6 (2001), 873.

3. Daniela J. Lamas, "'You're Dying,' I Told My Patient. I Wish I Hadn't," *New York Times*, October 6, 2021.

4. Arthur Dobrin, "Lying to a Dying Patient," *Psychology Today*, October 25, 2021.

32. THE VEIL OF IGNORANCE: RAWLS'S THEORY OF JUSTICE

1. Thomas Hobbes, "On the Natural Condition of Mankind as Concerning Their Felicity and Misery," chap. 13 in *Leviathan* (1588).

2. John Rawls, *A Theory of Justice* (Harvard University Press, 1971).

33. SHOW AND (DON'T) TELL: ZERO KNOWLEDGE PROOF

1. Richard W. Feldmann, Jr. "The Cardano-Tartaglia Dispute," *The Mathematics Teacher* 54, no. 3 (1961).

34. LIMITED LIABILITY: THE CORPORATE VEIL

1. Butler Murray, address at the one hundred forty-third Annual Banquet of the Chamber of Commerce of the State of New York, November 16, 1911.

35. MOVE TO STRIKE: UNRINGING THE BELL

1. State v. Rader, 62 Or. 37, 124 P. 195 (1912). Oregon Supreme Court, 62 Or. 37, 124 P. 195, https://cite.case.law/or/62/37/#p40.

2. Dunn v. United States, 307 F. 2d 883—Court of Appeals, 5th Circuit 1962, https://scholar.google.co.il/scholar_case?case=1039260860826380 9075&hl=en&as_sdt=6&as_vis=1&oi=scholarr#r[4].

3. Steven C. Day, "Getting More Than You Asked For: The Nonresponsive Answer," *Litigation* 14, no. 1 (1987): 18–20, 53–54.

36. HOT STOCK TIPS: INSIDER TRADING

1. SEC v. Switzer, 590 F. Supp. 756 (W.D. Okla. 1984)
2. Reuters, "Sanford Weill's Wife Named as Unwitting Stock Tipper," *New York Times*, May 18, 1990, https://www.nytimes.com/1990/05/18 /business/sanford-weill-s-wife-named-as-unwitting-stock-tipper.html
3. Associated Press, "S.E.C. Says Marriage Counselor Profited on Insider Information," *New York Times*, December 14, 1995, https://www.nytimes .com/1995/12/14/business/sec-says-marriage-counselor-profited-on -inside-information.html

37. CARNAL KNOWLEDGE: ADAM AND EVE

1. Martin Ingram, *Carnal Knowledge: Regulating Sex in England,1470–1600* (Cambridge University Press, 2017), 297.

38. IGNORE IRRELEVANT ALTERNATIVES: THE THEORY OF GAMES

1. Maurice Allais, "Le comportement de l'homme rationnel devant le risque: critique des postulats et axiomes de l'école Américaine," *Econometrica* 21, no. 4 (1953): 503–46.

40. POLLS ARE LIKE PERFUME . . . THE BANDWAGON EFFECT

1. Herbert. A. Simon, "Bandwagon and Underdog Effects and the Possibility of Election Predictions," *Public Opinion Quarterly* 18, no. 3 (1954) 245.
2. WAPOR, "Freedom to Publish Opinion Polls 2023," https://wapor.org /publications/freedom-to-publish-opinion-polls/.

42. CURSED BE HE: SPINOZA'S THESES

1. Steven Nadler, "Baruch Spinoza," *Stanford Encyclopedia of Philosophy*, ed. Edward N. Zalta and Uri Nodelman (Stanford University, Spring 2024).

2. Benedict de Spinoza, *The Ethics*, trans. R. H. M. Elwes (Gutenberg EBook, 2009), https://www.gutenberg.org/files/3800/3800-h/3800-h.htm.

3. Albert Einstein, *"Ich glaube an Spinozas Gott, der sich in gesetzlicher Harmonie des Seienden offenbart,"* quoted in German in *New York Times*, April 25, 1929.

45. KABBALAH:
ONLY MEN OVER FORTY

1. Gershom Scholem, *Major Trends in Jewish Mysticism* (Schocken, 1941), 34.

46. MURDER, SUICIDE, OR THE GODS' WRATH? THE SQUARE ROOT OF 2

1. Anthony Gottlieb, *The Dream of Reason: A History of Philosophy from the Greeks to the Renaissance* (Norton, 2001), 25.

2. Walter Burkert, *Lore & Science in Ancient Pythagoreanism* (Harvard University Press, 1972), 455.

47. *"EPPUR SI MUOVE!"*
THE TRIAL OF GALILEO

1. "Trial of Galileo," http://law2.umkc.edu/faculty/projects/ftrials/galileo/recantation.html.

48. DIAMONDS ARE A GIRL'S BEST FRIEND: A MINERAL CONSISTING OF PURE CARBON

1. James E. Shigley, "HPHT and CVD Diamond Growth Processes: Making Lab-Grown Diamonds" (GIA, July 25, 2016), https://www.gia.edu/hpht-and-cvd-diamond-growth-processes.

2. Thorstein Veblen, *The Theory of the Leisure Class: An Economic Study in the Evolution of Institutions* (Macmillan, 1899).

50. "NON, JE NE REGRETTE RIEN"
ANTICIPATED REGRET

1. Gerd Gigerenzer and Rocio Garcia-Retamero, "Cassandra's Regret: The Psychology of Not Wanting to Know," *Psychological Review* 124, no. 2 (2017): 186.

53. OLDIES PREFER IGNORANCE:
TWENTY-SOMETHINGS VS. OVER-EIGHTIES

1. Ralph Hertwig, Jan K. Woike, and Jürgen Schupp, "Age Differences in Deliberate Ignorance," *Psychology and Aging* 36, no. 4 (2021): 407–14.

54. PROTECTING THE SECOND AMENDMENT:
THE DICKEY AND TIAHART AMENDMENTS

1. Centers for Disease Control, "Fast Facts: Firearm Violence and Injury Prevention," (2023) stacks.cdc.gov/view/cdc/136202/cdc_136202_DS1 .pdf
2. Arthur. L. Kellerman et al., "Gun Ownership as a Risk Factor for Homicide in the Home," *New England Journal of Medicine* 329 (1993): 1084–91.
3. Gary Kleck and Marc Gertz, "Armed Resistance to Crime: The Prevalence and Nature of Self-Defense with a Gun," *Journal of Criminal Law and Criminology* 86, no. 1 (1995): 150–87.

55. IGNORING HIGH-RISK EVENTS:
DISASTER INSURANCE

1. Kenneth Arrow, preface to *Disaster Insurance Protection: Public Policy Lessons*, by Howard Kunreuther (Wiley-Interscience, 1978).

56. LEGISLATING MATHEMATICAL TRUTH:
Π = 3.2?

1. As quoted in Arthur E. Hallerberg, "Indiana's Squared Circle," *Mathematics Magazine* 50, no. 3 (1977): 136–40.

57. TO TEST OR NOT TO TEST?
PROSTATE-SPECIFIC ANTIGENS

1. National Cancer Institute, "Prostate-Specific Antigen (PSA) Test," updated January 31, 2025, https://www.cancer.gov/types/prostate/psa -fact-sheet#how-are-researchers-trying-to-improve-the-psa-test.
2. Paul F. Pinsky et al., "Extended Mortality Results for Prostate Cancer Screening in the PLCO Trial with Median Follow-Up of 15 Years," *Cancer* 123, no. 4 (2017): 592–99.

58. BEWARE: VIOLENCE AHEAD!
TRIGGER WARNING

1. American Association of University Professors, "Academic Freedom and Tenure," as cited in Edward J. Graham, "Mandated Trigger Warnings Threaten Academic Freedom," *Academe*, November–December 2014, https://www.aaup.org/academe/issues/100-4/mandated-trigger -warnings-threaten-academic-freedom.
2. John Ellison, "Dear Class of 2020 Student," University of Chicago, https://news.uchicago.edu/sites/default/files/attachments/Dear _Class_of_2020_Students.pdf.

60. EXPERTS, SHMEXPERTS: DENIAL OF
EXPERTISE, DISMISSAL OF ELITES

1. Tom Nichols, "The Death of Expertise," *The Federalist*, January 17, 2014, https://thefederalist.com/2014/01/17/the-death-of-expertise/.

INDEX OF NAMES

GPSR Authorized Representative: Easy Access System Europe, Mustamäe tee
50, 10621 Tallinn, Estonia, gpsr.requests@easproject.com